ATTRACTION
Secrets of Gravity

By
Reuven Nir

CONTENTO**NOW**

Attraction
Secrets of Gravity
Reuven Nir

First published in Hebrew 2000
Producers: ContentoNow
Translation to English: Shelley Raitskin
Cover Design: Benjie Herskowitz
Illustrations: Reuven Nir

ISBN: 9-789-655-506-18-1

International distributor:
ContentoNow
3 Habarzel St. 6971007 Tel-Aviv, Israel
netanel@contentonow.com
www.ContentoNow.com

*Dedicated with love
to the memory of my parents,
Pnina and Yeshayahu Nir,
among the founders of Kibbutz Merchavia.*

TABLE OF CONTENTS

INTRODUCTION

Prof. Chemi Ben Noon

*A scientific theory must be examined in light of the chances
of it having continued theoretical research.*

*Future research would deem it useful, whilst the absence of
research would deem it useless ...*

*A*ttraction – Secrets of Gravity deals with the main questions
that have perplexed humanity throughout the generations, such
as: What exists? What is a mass? What is time? What is force?
Its answers to the questions of "why do apples fall down?" and
"is the horse pulling the cart?" are different from those of Isaac
Newton. "The Attraction Force of Gravity" is mistaken in the
way it interprets the observations. It is not **attraction** that is
taking place, but rather **repulsion.** This book offers a fascinating
explanation to the unanswered questions remaining from the
theory of gravity, the theory of relativity and quantum theory. It
does not deal with improving that which exists, but returns to the
basic questions of the structure and motion of matter. The author
refers to the language, interpretation, understanding and limits
of human perception and thought, and tries to provide a simple,
coherent explanation to physical phenomena that are currently
explained differently.

Fracticles, which are probably the "existing matter" that Democritus meant when he described the concept of "A-tomos" (the indivisible), or what Leibniz meant when speaking of the "windowless monads." They might also be the answer to the ontological question, "what exists?"

Fracticles are the elementary particles of matter that move and pass, usually with ease, through bodies (which are merely "… local vortexes or an accumulation of disturbances"). What we term 'mass' is just a whirling deceleration of fracticles, which collide as they change themselves and others. Since they are several orders of magnitude smaller than what are currently considered to be the basic building blocks of matter, we do not have the means to prove their existence. However, it is possible to offer indirect evidence to establish their existence, and this is presented in the book.

Alongside the ontology, the author also examines the epistemology – the fracticles that enable us to grasp the world on two levels. The first level is conventional-cultural, which portrays a world view that is perceived through our senses, and continues through the nerves to the brain, where it is processed and "presented to us." The other is the abstract level, which might be a true explanation, but due to the limits of our understanding and thought, it enables us to discuss it only as a "possibility."

It would appear that Democritus, who believed that the world has motion, would celebrate his late triumph over Parmenides, who claimed that the world is one, and thus has no motion. In contrast, the author perceives the world to be a massive universal flood of fracticles that are invisible, are several orders of magnitude smaller than the known particles, and are built in a way that has still not been presented. Their movement is eternal and retains a regular velocity, which is the speed of light. It is only a disruption to that motion that might be perceived by us. A disruption occurs when there is an accumulation of "disruptive fracticle vortexes" such as you, myself and the planets. We only see, feel and measure their velocity, which is halted for a short periods of time. Speaking of **time**, that is just a creation of our culture rather than a physical phenomenon or resource.

The main role of *Attraction – Secrets of Gravity*, which is carried out very successfully, is to undermine our basic concepts of existence and the universe. Still, the explanation given in the book to various phenomena is not only in accordance with physics, but it might even improve it through the discussion of the most fundamental cosmology and ontology. If the velocity of fracticles is borderline, then different phenomena, from "red shifts" to "black holes", are given such a simple explanation, that one might feel as though "that is it," here is a fascinating solution to issues that have been discussed for such a long time.

Even if we assume for a moment that all of the explanations in the book are mistaken, which would be difficult to prove (though that is not to say that the claims are not falsifiable according to Popper's criteria), I believe that this book provides a cognitive breakthrough, equal in its potency to the explanation offered by Einstein, as he presented the Theory of Relativity.

It is an enormous challenge to our whole outlook, our world view and all that surrounds us – against a conventional scientific world, one that is stable, comfortable and very well-funded, and which could be hostile to such a revolution.

Hence, we must not disregard *Attraction – Secrets of Gravity*! We must not decide that "it is all unfounded" just because today's scientific mainstream may think otherwise. We must address ideas, the claims and the examples, because they might bring about not only changes in scientific concepts, alongside interesting advancement and research, but even changes in cultural concepts. Our comfortable, convinced world accepted the Theory of Relativity and Quantum Theory. It did not get angry at Heisenberg's uncertainty principle, did not scream hysterically when a significant logical principle was taken away from it (the law of excluded middle) and did not cry when an "improved" Newton paradigm took over. I hope this example will enable the ideas of *Attraction – Secrets of Gravity* to seep in, and allow us to further advance the level of explanation and research.

Even if there is no truth in all of this[1], it remains a fascinating intellectual experiment that may take the interested reader away from the dominant paradigm and prove that one must not sink in the warm swamp of "suitable explanations" by mainstream scientists, which is potentially full of contradictions and faults, for which there are no proofs.

One must keep researching in depth and clarifying. One must keep searching. That is the role of the scientist and the philosopher: not to give up, until the truth is revealed, until our world view becomes clearer, or alternatively, until we know each moment that this picture may only be a very powerful, accepted lie. Those who refuse to live in a lie, whose spirit and minds are strong, may take the reins and continue from the point Reuven Nir has reached in *Attraction – Secrets of Gravity*.

As a scientific editor, I have attempted to present the author's opinion with the utmost clarity. That is what I aimed for, and I hope I have succeeded. The editor and I tried to preserve the authenticity and originality of the writing. Thus, we only added a

---

1   In his lecture, The Feynman Lectures on Physics (Wesley 1963, Volume 1, paragraphs 7-9, 7-10), the renowned physicist Feynman voiced his doubts regarding particles and repulsion as an explanation for gravity. He did not suggest any alternative explanation to Newton and Einstein's assumptions on what gravity is, or why the universe expands at an accelerated rate. Recently, there may have been other works done, both in Israel and around the world, which may have similar ideas. If that is the case, Reuven Nir will bravely and persuasively join in these discussions, with an adequate measure of skepticism (editors).

few footnotes, to help those interested readers who might like to expand on the topics discussed.

Apart from footnotes, there is additional information in the **Appendices**. There you will find a **Recommended Bibliography**, a **List of Notable People** with short summaries of their lives and writings, a **List of Terms**, with brief explanations and **References for the Excerpts**, which may lead to further, complementary reading.

Finally, it seems that *Attraction – Secrets of Gravity* joins a long, respected list of philosophy books that discuss vital questions, such as what is the universe, what is it composed of, and how can one explain physical phenomena simply, without contradictions and with fewer obstacles than the ones faced by philosophers in the past.

Chemi Ben Noon
Ramot Hashavim, 1999

PROF. CHEMI BEN NOON holds degrees in Mathematics, Law and has a PhD in Philosophy. He lectures at universities and other educational institutions. He founded the Kiddum Institute, is the Chairman of *Mad Science of Israel* and a Research Associate at the International Institute for Counter-Terrorism at the IDC in Herzliya. He also co-authored the books *Preparation for Psychometric Exams* and *Improving Thought Processes* and is the author of *Civil Rebellion, The Limits of Intelligence – A Discussion between Yeshayahu Leibowitz and Joseph Agassi*, and *The Civil Appeal*. He is the editor of *Discussions on the Philosophy of Science* on The Broadcast University (academic lectures on national radio) *The 20th Century*, and *Letters to My Sister*, and is the scientific editor of *Attraction – Secrets of Gravity*.

How the Manuscript Evolved

Reuven Nir

One summer night, when I was about four years old, I sat with my father, Yeshayahu, may he rest in peace, on the steps of my parents' home in Kibbutz Merchavia, looking at the stars shining in the sky and wondering what were those twinkling lights up above.

My father, who would later become a teacher of math and physics, explained to me that the stars were not torches hanging in the sky. They are distant worlds, the shape of a huge shining ball, most of which are bigger than the ball we live on, called "Earth" (at that same time, I learned that we live on a ball). Most of these stars are bigger than the sun, but they look small, due to their great distance from us.

When I was a bit older, my father gave me a map of the night sky and field binoculars, which made the stars look four times bigger than they were. In the dark, summer nights, I would go outside, lie down on the grass, and, using the map, I would try to identify the star groups in the sky and look between them for distant nebulas. At that time, there was still no clear distinction between a nebula and a galaxy.

At the age of 12, I read *The Stars in Their Courses* by James

Jeans, and *The Birth and Death of the Sun* by George Gamow, which affected my thinking in the following years.

I chose to combine illustrations from *The Stars in Their Courses* (which was written in 1931) on the inner cover of the first edition published in Hebrew in 2000. These illustrations may look somewhat primitive, compared with the amazing photos from deep space and the remarkable maps of the sky that today are accessible to all. It is a modest reminder of the sources from which people of my generation learned about the universe. Unfortunately they are not included in the translated English text.

I felt uncomfortable in physics class when I learned about the attraction force of gravitation , and this discomfort accompanied me later on as well, during my studies in London.

I did not find a satisfying explanation for the phenomenon of the attraction force of gravity in the books and articles that discuss perceiving the world around us, both physically and philosophically, by understanding the laws of nature.

In 1957, I established an explanation, or an alternative theory to the classic attraction force of gravity, as phrased by Newton. This explanation is based on fast fracticles, which cause **repulsion rather than attraction**.

At that time, I was busy with other issues, so writing about the subject was delayed to some far, undefined time in the future, when I would possibly have some spare time.

In 1990, I first typed out the manuscript on the computer and in early 1992, I sent the "first edition" to several readers, in order to get their feedback. I cherish and appreciate the copies of the book with the comments I received.

At the end of September 1994, I had a heart attack, after which I had a bypass operation. I feared that the elevator that took me down to the operating rooms in the hospital would be the last thing I would see in my life, if the operation failed. I had a vivid sense of the temporariness of life, and I also feared that ***Attraction – Secrets of Gravity*** would not be published in my lifetime or after that. My view of life immediately before the operation as a temporary, passing thing, was what pushed me to continue and complete the work.

Some of my best friends helped me with the task. These included my brother, Yitzhak Nir, an airline captain with El Al, who produced, edited and designed the book; my friends, Shalev Man, Dan Raz and the team at the creative company where I worked; Dr. Chemi Ben Noon, the scientific editor and writer of the introduction; Major General (Ret.) Herzl Bodinger and his wife Orit; and Prof. Shulamith Kreitler, who gave me their feedback.

I cannot overlook the direct "push" that I got from Ruthie Zohar and Dr. Gideon Ganot from the "governing council," of the Forum, on which I will not elaborate here...

Special thanks to my friend, Yael Golani, for her support and great patience.

In February 1995, I sent a revised edition to Keter Publishing House, as well as to Zmora-Bitan publishing house, Maariv publishing house and Yediot Books publishing house. The manuscript was transferred to their team, and thus, some of the ideas in the book already became public.

It was only in 1999 that Yoram Rose from Kinneret Publishing House agreed to publish ***Attraction – Secrets of Gravity***, for which I am grateful.

In the 1960's David Ben-Gurion, Israel's first prime minister, approached Samuel Sambursky and asked him to write an anthology on the development of the sciences and on their relation to philosophical approaches and the establishment of theories in science.

I have included some quotations from this important book, *The Evolution of Physical Thought – From the Pre-Socratics to the Quantum Physicists – An Anthology*, which was published in 1972, in several chapters in the book, to connect between perceptions of the past and the new ideas that I present in the book.

Other chapters have quotations from additional sources. I hope that by doing so, I have brought some nostalgic flavor to these chapters.

The Recommended Bibliography in the appendices at the end of the book does not refer to quotations that appear in the book itself, but recommends additional books or articles that are not discussed, yet have affected my way of thinking (including some views that I disagree with). These could enrich the minds and imaginations of interested readers.

Some of the quotations were added by the editor and the scientific editor.

The life story of a Polish girl named Maria Skłodowska, who arrived in Paris in 1891 to obtain an education, caught my imagination when I was a child. She lived in poverty in an attic, but finally married the French scientist, Pierre Curie. Through hard work and many difficulties, the two of them discovered the chemical elements polonium and radium as well as radioactivity, at a time when it seemed that there was nothing left to discover in the field of exact sciences.

In 1906, Pierre Curie was run over by a heavy horse-drawn cart. Maria, who continued to publish their research, would then become known as **Madame Curie**, gaining international fame. The part that describes "a horse pulling a cart" is dedicated to the memory of the Curies and to their persistent aspiration to discover the secrets of matter.

Attraction - Secrets of Gravity deals with qualitative thought, with barely any quantitative expressions of equations or calculations, apart from the section discussing the relation between the force of gravity or repulsion and distance (Chapter two: "Distance and the Force of Repulsion"). This section is accompanied by a geometrical description and illustrations, which I hope will help those who are not familiar with these fields. One can skip it, without worrying that it may affect the continuum of thought or the understanding of the following chapters.

The book was written for laymen, and for people of all ages who are interested in expanding their horizons, and does not require any prior knowledge of physics, math or the philosophy of science. Therefore, perhaps there may be more openness to new ideas among those who did not obtain a formal academic education.

The writing of *Attraction - Secrets of Gravity* took me a long time, and was conducted independently from work that may have been written about similar ideas, since the eighteenth century and until today. If that is indeed the case, I would be pleased if *Attraction - Secrets of Gravity* had a place among such work, and would contribute in clarifying the problems at the forefront of research today.

I would like to take this opportunity to also thank Meir Salomon, Danny Lasri, Dr. Rami Kalir, Chaim and Shlomit Aluma, Lina Arnon, Tal Becker, Dr. Shmuel Gordon, Yitzhak David, Ilana Edelman and the late Amos Vining, for reading my manuscript and sharing their thoughts with me. Thanks to Dr. Doron Mor from Kibbutz Merchavia, for information about the magnetic properties of parts of Basalt rocks on the Golan heights.

Thanks to everyone who encouraged me in my first steps and to those who read my manuscript or parts of it during the writing process, enlightening me with their questions, responses and comments, and so contributing to the book and its final format.

Reuven Nir
Kibbutz Merchavia,
Spring 1999

CHAPTER 1

- "Truth", or Interpretation?
- Language
- Why do Apples Fall?
- Is the Horse Pulling the Cart?
- Attraction

"Truth", or Interpretation?

"I set forth the reasons for which we may, generally speaking, doubt about all things and especially about material things, at least so long as we have no other foundations for the sciences than those which we have hitherto possessed. But although the utility of a Doubt which is so general does not at first appear, it is at the same time very great, inasmuch as it delivers us from every kind of prejudice, and sets out for us a very simple way by which the mind may detach itself from the senses."

RENE DESCARTES

"You do not have to love philosophy or believe in it. You must climb it like a ladder, and if it breaks - get another ladder."

WERNER ERHARD

There is no truth in this book!

All of the words, sentences and ideas within it may not be true and they may not represent facts or results of research or experiments that are "completely true."

One should, perhaps, refer to the new ideas presented in the book, alongside previous ideas, not as ideas that are true or false, "real" or fabricated, but as concepts that were created by us, humans. These concepts that we have created, that are a significant part of every language, may enable us to observe objects and movements in all that our senses receive.

For instance, we can see a tree with a background of the

18

sky, since the concept of "tree" enables us to observe the tree as a separate unit than the concept of "sky," rather than a mixture of leaves and sky or patches of green and blue, as seen by our eyes.

The concept of "tree" may enable us to group together the trunk, the branches and the leaves, which we see in a certain area, without including in it bits of sky (and maybe some birds too) that are seen under the leaves, even though we see the leaves and the sky as one continuum.

The concept of "view" enables the tree and the sky to belong to one unit that also includes other objects or groups of objects for which we have names, and thus it creates a new unit that can be termed "view."

A painting, which has green and brown patches, between which are patches of light blue, would look like "a tree with the sky in the background," since the concepts "tree" and "sky," which are etched in our memory, enable us to see the tree as a separate entity from the sky, rather than a mixture of different colored patches.

Just as one could see either patch in the patches of color, or essential parts of the tree or the view, thus one may also see the "truth" or "untruth" as linguistic concepts or parts of the language[2], rather than as real parts of our surrounding reality.

These concepts may be interpreted differently by different people, depending on conditions, outlooks, knowledge, thought and imagination, and do not necessarily represent an external reality.

To emphasize the concept of "interpretation" with regards to the "external reality," I will use the famous anecdote about the sculptor Michelangelo: Let us try in our imagination, just like Michelangelo, to "see" the statue with all its details, as if it is already

---

2   A most intriguing opinion regarding language is expressed in the book Human and Cosmic Thought, by Rudolf Steiner. Appendix: "Recommended Bibliography" (editors).

embedded in the marble, even before sculpting work has begun.

According to Michelangelo, the statue is already there, deep in the rock, waiting to be exposed. The sculptor has just to remove the "excess" marble, and the new statue will be revealed to all...

The marble rock has many statues "embedded" in it. Perhaps a different sculptor will remove other shards of marble and expose another statue?

A person passing by the statue may not notice it, or may stop, full of admiration for the magnificent work of art. A scientist may only see in the statue a collection of atoms and molecules of different kinds that compose something that is generally known as "a lump of marble."

We are the ones who call the lump of marble "a statue," according to our perception at the time, or the way we interpret what we see. We are the ones who attribute to that "cluster of atoms and molecules," the "shape," the "beauty" or "the meaning," and generally, its existence as a "statue."

Perhaps everything we see, hear, taste, smell or experience in any possible way, is a particular interpretation of "things" which are around us, but not necessarily "the reality" or the things themselves.[3]

If our senses are the only source to provide us with information on what occurs around us, then the way our brain interprets what is received by our senses is reality for us![4]

If there is information that our senses or our equipment are not able to receive, that information does not exist for us! It may exist in reality, but not in a form that is accessible to us.

---

3   An idea that is similar to the ontological "being" argument, which, according to philosophers such as Kant and others, we will never be able to achieve (Editors).

4   Those wishing to expand on these subjects may also read The Brain, The Final Frontier, by Richard M. Restak. (Appendix: "Recommended Bibliography" editors).

If we imagine a world where all people are born blind, we can imagine that in such a world, there is no place for light. Furthermore, it will not "have" a moon or stars in it, which may be viewable by one member or another of that human species.

This fact does not "bother" the moon and the stars at all, and does not cancel their existence.

A blind scientist, who builds a telescope in a world of blind people, may be considered to be a man who has lost his mind, rather than a researcher trying to investigate the meaning of phenomena in the universe.

Without a sense of vision, the whole world of concepts that depends on this sense is canceled; all the words that relate to light, color, shape and sight disappear, and all the equipment related to light and vision become worthless.

In a world where all people are deaf, it may be considered pointless to listen to music, rather than an art that brings pleasure to both the musicians and the audience. In "the world of the deaf," there is no place for a violin, a piano or a CD player, and a conductor of an orchestra would quite possibly appear to be an inconceivable phenomenon.

The lack of a sense denies us information that comes from some external reality and "pushes" us to a specific interpretation of our world.

If we had a new sense, which had been previously unknown, we would possibly perceive the same reality in a different way, and scientists would build different equipment, based on new principles, and our concepts of the world around us would expand and change considerably.[5]

A multiplicity of senses would perhaps offer us a clearer,

---

5   Do bats, which find their way in dark caves by using their high-frequency shrieks, as well as gliding migratory birds, have a sense of direction and orientation that we lack?(editors).

more comprehensive and more reliable perception of reality, just as a "lack of senses" can only present a fragmented, unreliable and inexplicable reality in many cases.

Could it be, that the "exact sciences," as they are known to us, are not exact, or that, they do not present a complete portrayal of external reality?

Perhaps sciences offer merely a narrow, one-sided view of the universe, rather than a complete, authentic description of what occurs around us, as we have only five senses, rather than a hundred?

Those who grew up studying clear "physical truths," proven with mathematical equations, may find it difficult to accept a claim stating the falsehood of these "truths," unless they adopt an approach whereby there is no truth or untruth, but various interpretations of reality.

Does a scientific approach have no interpretation?

This book invites you, the reader, to take an active part in developing a new way of thinking, which may differ from the approaches to which you are accustomed.

Not all the chapters in the book are theoretically "closed," which is why I would prefer to call them "openings." Through them, one can enter into a new, unknown areas, spaces or structures, even if all their building blocks are not in place yet...

Where do these openings lead? We still do not know.

Should we take a risk and pass through one of them, at least?

LANGUAGE

"As the goal of science is to augment and order our experience, every analysis of the conditions of human knowledge must rest on considerations of the character and scope of our means of communication. Our basis is, of course, the language developed for orientation in our surroundings and for the organization of human communities. However the increase of experience has raised questions as to the sufficiency of the concepts and ideas incorporated in daily language. Because of the relative simplicity of physical problems, they are especially suited to investigate the use of our means of communication indeed..."

NIELS BOHR

"Equations are necessary if you are doing accounting but they are the boring part of mathematics. Most of the interesting ideas can be conveyed by words or pictures."

STEPHEN HAWKING

It is widely thought that matter is composed of atoms, and that an atom consists of a nucleus bearing a positive electric charge, surrounded by electrons bearing a negative electric charge.

What does the concept of "negative" or "positive" electric charge mean to us?

Perhaps the "positive" or "negative" concepts do not represent something clear, but a property which we could not describe in any other way, as we do not have a sense which distinguishes between "positive" and "negative" in the physical meaning.

Each one of the "elementary" atom particles, as we "understand" or interpret our observation, might also be composed of smaller particles, which in turn may also be composed of even smaller particles, and so on…

How small can the "elementary" particles of which "matter" is composed be? Is there a limit to how small the scale can be?

If matter is composed of small particles, which are made of even smaller particles, then what is motion, that phenomenon where "matter" or "clusters of particles" move from one place to another?

Does the **place** itself have an independent existence, which does not depend on those particles?

We assume that motion includes within it what we call "time," and also "velocity" or "energy" or "momentum." Do these words express an independent "physical meaning," namely, the **things themselves**, or merely the interpretation of the experiences of our senses?

Could there be **a different interpretation** to those concepts?

Could it be that when we examine concepts, which seem to us to be known, "real," clear and tested, such as **motion, time, velocity, matter, energy,** or **force** and **mass**, they may lose their definite element? That they will cease to be real, proven or unquestionable, and we will have to draw on theories and images taken from our spoken language, on which there is a wide agreement?

Is it not language that creates the concepts and enables us to agree on them, and thus defines their existence?

Perhaps all physical concepts do not exist in nature, as they were created, but exist in language, as created by man.

Perhaps what causes a certain claim to be true or real is the

agreement many people have about its validity or truthfulness, rather than a particular, intrinsic property it has.

Perhaps the cluster of atoms and molecules in the carved lump of marble seems to us like a statue of David or Moses, since there is an agreement, which matches the common image in the imagination of many peoples, who have read the bible or studied about King David or Moses.

Maybe matter is known as "matter," since there is a wide agreement to term it so, and even its exact nature is not entirely clear.

Perhaps motion, energy, time, mass and force are so, because there are words in our language that describe them so, and not because that is the way they were created or have always existed.

Which world is truer: that which has **four** elements (fire, water, air and earth) as considered by the Greeks in ancient Greece, or that which has more than **a hundred** chemical elements, as we know nowadays?

Maybe both viewpoints are equally "correct" - perhaps they are both possible interpretations of the same world? In both outlooks, a theory was built, through a process called "thought," based on fractions of information which we receive through our limited number of senses, from some reality that exists **outside of us**.

Could it be that things in which we believe, or that we consider to be true, are no truer than the things in which we do not believe, or which we consider to be untrue?

Is it possible that it is only our interpretation, together with the agreement of those around us, which makes these perceptions true, while the interpretation of others casts doubt on them?

Doubt might be the preferable interpretation, as it does not close our mind to other interpretations or ways of thinking.

The concept of "God" is vivid and logical to a religious person, and intangible and at times illogical to a secular person. They both live in the same universe, have the same senses, but they interpret the universe and its laws in different ways. Thus they "see" the universe in completely different ways.

Do all people see the same world in their eyes or hear the same sounds in their ears?

It may be that every person sees and hears images and sounds differently from others, and the usual way to describe what he experienced is through language: in speech, writing, drawing, in sound or in movement.

Perhaps language expresses something that each person processes with their own "basic software," based on a limited number of words, expressions, sentences and ideas.

Even the simplest words, sentences, expressions and ideas are understood slightly differently by each individual. However, since the variance is not so great, we "manage" with each other, more or less.

For instance, when hearing the word "airplane," a person may envision a padded seat in an airliner; a pilot might envision the cockpit, his equipment, or an occurrence that happened during a flight; and a little child might imagine a big bird flying in the sky.

Each one of the three "sees" a different image, which matches his experiences, his pool of concepts, and his inventiveness, called "imagination."

We can assume, interpret or "understand" what each of the three saw, based on the description and explanations they give, and based on our experiences and our pool of concepts and imagination.

"Contemporary physics," which describes numerous phenomena, tries to do so while finding a narrow common denominator, for as many as possible observed facts.

Does contemporary physics actually handle facts, or merely **words** or **concepts** that describe them?

Could there be direct access to facts without language that describes them?

When we look at the sky on a dark night, and see thousands of stars twinkling in the sky, do we see a uniform image of stars, as they are today?

The light of the "fixed stars" that are closest to us (which are beyond our solar system) arrives to earth after approximately four years. The light of other fixed stars arrives to us after hundreds, thousands or even millions or billions of years. Thus, the image of the sky that we see on a dark night is composed of the light of stars as they **were** all those years ago! It is possible that some of the stars we see do not exist anymore. Yet nonetheless, their light still shines in the night sky.

We see a particular, uniform image of the sky, and know that it does not correctly represent the existence of stars today.

Even contemporary physics does not present one unified, updated and correct image of reality. It offers a large number of results of experiments and interpretations of those experiments, which were carried out in different periods, by many different researchers.

Scientific theories are updated and change from time to time, so perhaps there is no absolute truth in physics as well?

What was true once is not true today. Furthermore, what is true today will only remain true until new facts are revealed to contradict "what is true," or until a new theory better explains the observations.

Perhaps what we term an explanation, an understanding or a scientific theory, are no more than a particular, subjective interpretation of the observations.

Could it be that the theory of gravity, the attraction force that masses in the universe exert over each other, is no more than one possible explanation of the observations?

Could it be that the theory of relativity and the motion of bodies within "curved space-time" are no more than another theory or another possible interpretation of the observations, as are the rest of the theories in science?

It seems that a mathematical equation does not add to validity of the theories, since math is just another language, among other languages. Math needs to be studied, just like other languages. Letters or signs are explained by using words and sentences from the spoken languages. Thus, perhaps we can say that math is "an accurate translation" of the spoken languages, but it is not "the original."

Just like other languages, math serves as a tool with which we "analyze" or interpret the observed facts or contexts and the relations between them as we understand them, but it also does not express the facts themselves.[6]

A new theory may perhaps predict new facts, and offer better explanations to the known facts, and to the connections between them. However, as we accept one theory, we close the way to other theories, and so perhaps deny ourselves the possibility of understanding reality more clearly.

Every new interpretation contributes greatly to the understanding of reality and lightens it from a different angle. However, at the same time, it blocks other possibilities that may,

---

6   A mathematical equation may help to sift through internal logical failures from new theories and ideas (editors).

perhaps, contribute more. When a certain interpretation of the research of the universe ceases to advance us, should we not search for a new interpretation that would open new possibilities, which were blocked by the previous interpretations?

Why Do Apples Fall?

"Hitherto, we have explained the phenomena of the heavens and of our sea by the force of gravity, but have not yet assigned the cause of this force. This is certain, that it must proceed from a cause that penetrates to the very centers of the sun and planets, without suffering the least diminution of its force, that operates not according to the quantity of the surfaces of the particles upon which it acts (as mechanical causes used to do) but according to the quantity of the solid matter which they contain, and propagates its virtue on all sides to immense distances, decreasing always in the raising to a power of the distances..."

ISAAC NEWTON

"Every weight tends to fall towards the center by the shortest possible way."

LEONARDO DA VINCI

If we could follow Sir Isaac Newton's thoughts, as he saw an apple fall down out of a tree, perhaps we could phrase these thoughts simplistically thusly:

What caused the apple to fall down out of the tree?

Or, in more detail, and using different terminology:

What caused the apple to leave the tree and move, with increasing velocity, towards the center of the earth, until it hit the earth and stopped?

Furthermore, the answer to the question:

It seems that the earth is exerting some sort of force on the apple and pulling it to its center!

"This pulling force, called gravity," which operates in a straight line between the center of the apple and the center of the earth, must have caused the apple to detach itself from the tree and move, with increasing velocity, towards the center of the earth…

Newton termed this force "a central force," since the motion of the apple (and other bodies) starts at one point, and continues in a straight line through the air (or any other medium) towards the center of the earth.

Newton assumed that the apple also "pulls" the earth with the same force. And so, the two bodies move towards each other.

The distances that the apple and the earth pass in their movement toward each other are inversely proportional to their mass.

Namely, the apple, which has a small mass, passes a greater distance towards the center of the earth, and the earth, which has an immeasurably larger mass than that of the apple, passes a miniscule distance towards the apple.

The quantitative equation of that "attraction force" or "central force" or "gravitation," as phrased by Newton, was widely accepted in the generations that followed Newton, and, like any explanation that is widely agreed upon, greatly influenced the way most people thought.

The term "**attraction force**" was probably widely accepted by the public more thanks to the introduction written by Roger Cotes to Newton's book *Mathematical Principles of Natural Philosophy* [7] published in 1687, than to the writings of Newton himself.

---

7   From Latin: *Philosophiae Naturalis Principia Mathematica*

It may be that the effect this theory has over us today is so strong, that we tend to see it more as a proven fact and less as one possible interpretation of the observable results.

Perhaps we tend to add the qualitative explanation of **"why do bodies fall"** to the quantitative description of **"how much do bodies fall,"** in terms of direction, acceleration and velocity, and assume that if the quantitative description offers good results, necessarily the **qualitative** explanation that accompanies it must be true as well.[8]

Perhaps the assumption about the link between the observable or calculated results and the explanation to the question **"why does it happen,"** is somewhat hurried and not necessarily true, or at least not the only possible interpretation?

The quantitative formulation of Newton's Laws of Motion had a welcome influence on the way we understand the orbit of bodies that are close to earth and of distant bodies, such as the moon, the planets, the sun, the stars and the galaxies.

However, when the question "why do bodies move as they seem to move" was explained by the answer of **"attraction force,"** it seemed to have a "chilling effect" on the way many people thought in the generations that lived following Newton.

It is only very seldom that we stop to ask: **"Is attraction force indeed a real thing?"**

Does this concept derive from reality, or does it simply originate from an interpretation that has been accepted by many? Could it be that we accepted this theory without sufficient criticism, or at least without taking heed of our reservations or doubts?

---

8   Newton claimed on the one hand, that universal gravitation is based on mathematical derivations; while on the other hand, he wrote that God was the reason for gravity. He publicly denied his theological approach, so as not to question his distinguished position as a scientist and a mathematician.

Einstein's special and general relativity theories developed from the background offered by Newton's theory. They expand on the subject, and offer explanations to observations that Newton's work could not explain, such as the movement of Mercury around the sun, or the scale of curvature of the movement of light rays near massive bodies.

A popular description that Einstein gave to the existence of **"curved space-time"** as a possible explanation for gravity, visualizes such a space as a stretched sheet of elastic rubber, with a stone placed in the center of it, which creates an indentation in the sheet. Bodies placed on this sheet of elastic rubber, which simulates "curved space-time," will move towards the indentation that was created, that is they will deviate from moving in a straight line, and will move according to the curvature of the "sheet of elastic rubber," which represents "curved space-time." These bodies will act as though they are affected by an attraction force of gravity, and the evidence for that is their deviation from moving in a straight line, as would be predicted by the principle of inertia.

Would this not explain the motion of bodies, as "containing" or "comprising of" the existence of some force of gravity, without which the bodies placed on the sheet would have stayed in their places, despite the curvature of curved space-time?

Without gravity, the stone would have no "weight," and would not create any indentation in the sheet of elastic rubber, and thus no object would roll down to that point.

Can curved space-time change the motion of a body without creating some "material barrier" which would interrupt the motion of the body in a straight line?

Does the concept of "curved space-time" not **include** in it areas that are permissible and non-permissible for motion and leave open the question of what is the factor that "prohibits" or "allows" any motion in these areas of curved space-time?

It is very difficult to accept this simplistic explanation of "curved space-time" as a possible explanation or as the only explanation of gravity. This is because the concept of curved space-time already contains some sort of gravity which restricts the movement of objects in space to defined routes.

If curved space-time were made of jelly, for instance, rather than of elastic rubber, then the objects may have moved about in straight lines, despite its curvature, because the areas which were passable and non-passable would be less firm or less defined.

Newton and Einstein's pioneering work greatly advanced the development of science, and they can be seen as a powerful contribution to all the other theories and explanations that followed. However, they do not provide evidence that earth does **pull** the apple or any other object near it, or that **curved space-time becomes curved** as a result of there being a mass in it.

The famous experiment conducted by Sir Henry Cavendish in 1798 to examine whether there is an attraction force between every two masses, does not offer evidence of attraction, but it does provide evidence that the two masses **moved closer to each other.**[9]

Cavendish's experiment, conducted just over a hundred years after Newton published his work, confirms Newton's observations and shows that close masses do **move towards each other.** However, this experiment is not enough to answer the question - what **causes** this movement?

---

9   In this experiment, two small balls were hung at the two ends of a light, stiff rod, which was hanging on a thin glass thread tied to its center. When two larger balls were brought close to the small balls, the small balls moved towards the large balls and caused the glass thread to twist. A mirror was attached to the glass thread, onto which a narrow light beam was projected. The twist in the glass thread projected the reflection of the light beam to a graduated scale and thus the miniscule movement of the masses was converted into a significant and measurable deviation from their original state.

Could we assume, following Newton's work, that the force of gravity, as we understand it, or imagine it in our minds, is a sort of property which accompanies the mass, but is not the mass itself?

Could it be that this "property," the existence of which can be deduced from the movement of bodies close to the mass, is "embedded" in the mass, and moves with it wherever it goes?

This property, "attraction force," seems to start from the mass itself and reach the ends of the universe, creating a sort of ball-like halo of attraction, which weakens with distance, but never disappears or ends completely.

Can we accept that in the "real" world, a mass will move from one place to another, and at the distant edge of the universe, thousands or millions of light-years away, its halo (attraction) will move at exactly the same time and to the same extent?

What is that "thing" that moves alongside the mass, from its center to "the ends of the universe?"

What is the thing that we term "property" or "weight" or "gravity," which we attribute to bodies, but is not the bodies themselves?

Is there another way to explain the movement of bodies apart from attraction, a way that our way of thinking hides from us?

Is it possible to **understand** gravity-force, and not just to describe its results quantitatively?

Does gravity force or attraction force exist in reality, in the world of matter and motion, or just in our way of thinking and in the interpretation we give to phenomena that we come across?

Does it take time for gravity to reach from one body to another, or is the action immediate, regardless of the distance between the two bodies?

In order to illustrate this last question, let us imagine that the sun, which the earth has already revolved around in an almost circular movement for billions of years, will suddenly and instantly evaporate and disappear, leaving no material trace in what we term "the solar system."

If we put aside the unpleasant implications of this cosmic event would have on life in our world, we could wonder: Would earth continue to orbit for a while in an almost circular motion around the place where our sun used to be, or, would it **instantly** change its movement into linear motion at a constant speed in a blink of an eye, as if there never was a sun?

Could immediate action occur in the world of matter and motion?

If so, how would earth, which is a hundred and fifty million kilometers away from the sun, "know" instantly that the sun disappeared and that it must cease its almost circular movement and transition into a linear movement at a constant speed?

If we assume, contrary to Newton's theory, that "attraction force" expands in space at the speed of light, it may well be that earth would still continue its movement around where the sun used to be (before it disappeared) for another eight minutes (the time required for sun light to reach earth), before transitioning into a linear movement and disappearing somewhere in the dark expanse of the universe.

If there is an attraction force, and it acts between all bodies in "a finite universe," that is, if every celestial body in our universe **pulls** all other celestial bodies, why does the universe not contract, as one could expect based on Newton's theory, but on the contrary, the universe expands, as we learn from the "red shift" (discovered by the astronomer Edwin Hubble) of starlight coming from distant galaxies?

Why is the expansion of the universe, as observed in the past few years[10] **accelerated, rather than slowed down,** or at least not maintaining a more or less regular speed, as one could perhaps predict, based on the "Big Bang" theory?

If attraction force originates from atoms of matter, what is that "thing" created by the atom, which makes it **leave the atom** and affect distant bodies, even if there is no medium between them? How can one understand the concept of "attraction force" in physical terms?

The theory of attraction force has several limitations, such as the immediacy of its effect, its action across infinite distances with no medium between the "pulling" bodies, and the inability to examine or to understand the **mechanism** of that attraction.

As of today, we still have not found a particle or a wave which is in some way responsible for the "attraction" or "gravity" phenomenon, and so perhaps we should wonder whether we should doubt their existence.

The observed facts teach us that bodies **fall**, or **move towards each other,** but there is room to doubt the explanation claiming that this phenomenon is caused by an abstract factor termed **"an attraction force,"** or **"a gravitational force,"** or **"gravity"** or **"curved space-time"** which leaves no traces in the world of matter and motion.

Perhaps the search for the **graviton,** that "elementary particle" which was never found, that scientists search for, which they consider to possibly be the **"carrier of attraction force,"** will not help us much in understanding that phenomenon.

If **gravitons** arrive to the Earth from the sun, and Earth

---

10 Those wishing to expand on this subject may look at the work of the groups of Richard Ellis and Avishai Dekel in the Appendix: List of Notable People (editors).

serves as a restriction to their motion, they would push it away from the sun, and **not towards the sun,** as we think gravity force does.

It seems that "gravitational waves," or "gravitons," which leave one body and move towards another body, must **push** that body when they hit it and **distance** both bodies from each other (just as air molecules in the shock waves of bombs push and distance the bodies they hit and do not bring them closer to the source of the explosion).

Perhaps if the "gravitons" or "gravitational waves" are discovered in the future, they will repel masses from each other and not cause two distant masses to move towards each other. Perhaps in our universe, there are no **pulling** "gravitons" or "gravitational waves" that **pull** and do not **repel.**

It could very well be that we live in a universe **with no attraction force.**

If indeed we do live in a universe with no attraction force, **why, darn it, do apples fall down?**

Is the Horse Pulling the Cart?

"If a horse draws a stone tied to a rope, the horse (if I may so say) will be equally drawn back towards the stone: for the distended rope, by the same endeavor to relax or unbend itself, will draw the horse as much towards the stone as it does the stone towards the horse, and will obstruct the progress of the one as much as it advances that of the other."

Isaac Newton

Some of us may remember the old physics books, which showed a tired horse pulling a heavily loaded cart, where it said: "The horse **pulls** the cart, thus it exerts **a force** over it!"

What does the horse exert over the cart that causes it to move behind it?

If we get closer to the horse harnessed to the cart, we will see that the horse harness or collar, over which the horse exerts force, is on the horse's neck and chest, and not on its rear. Thus, the horse **pushes** the cart with its chest, and does **not pull** it, as we were taught to think. The horse pushes the harness collar, which transfers the force via the harness to two metal links, attached to metal rings, which are attached to the body of the cart.

If we get closer and check what the balance of forces is between each pairs of links and rings, we will see that the links **push** the rings from their inner side.

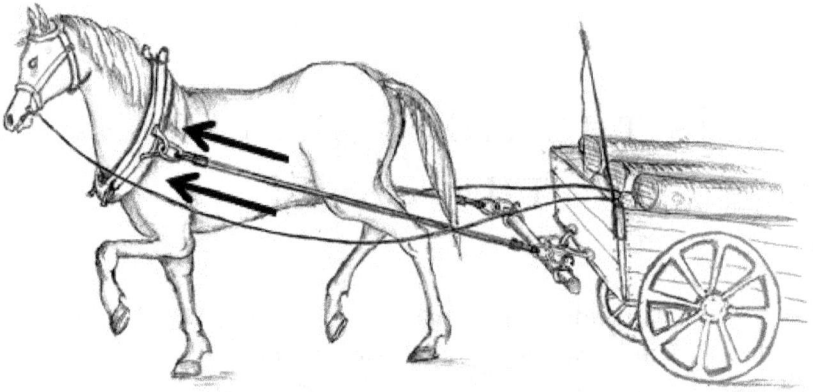

Illustration No. 1

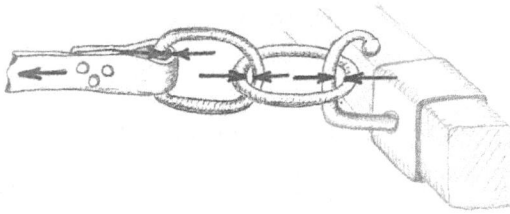

An enlargement of the area of forces exerted between the
rings and the links

Perhaps "a mechanism" similar to the one which operates between the links and the rings and which pushes them from their inner side, also functions between the atoms and the molecules of which the harness is composed of, and that this mechanism carries out the action throughout.

The image of links and rings does not completely match what we know today of the connections between atoms and molecules, but neither does it contradict it, and it might even open us up to a new way of thinking.

It might be easy for us to think in terms of the horse pulling the cart, because the horse is **in front of** the cart, and the cart trails **behind** it.

Many expressions in our language visualize different situations for the horse pulling the cart, or trying to pull it out of the mud, just as other expressions describe the rising or setting sun.

Since the days of Copernicus, we have recognized the fact that the rotation of the earth around itself creates the illusion that the sun "shines" in the east, "rises up" in the sky and finally "sets" in the west, in relation to the motionless earth.

Astronauts in a spaceship orbiting the earth can see many sunrises and sunsets in 24 hours, whilst their colleagues on earth will "only" see one sunrise and one sunset in that same period of time.

Just like the sun "rising" or "setting," the horse that is harnessed to the cart in the **front** creates the **illusion** that it is pulling it.

A horse harnessed to a cart in the **back**, with shafts connected to the **front** part of the cart, causes the cart to move forward, and it is clear to us that the horse is **pushing** the cart.

A horse harnessed **behind** the cart will **push** the cart, **exactly the same** as the horse which is harnessed in front of the cart. In both cases, the same action will take place, namely, pushing, and not pulling, as we were accustomed to think.

Pressure, or Vacuum?

In modern milking centers, cows are milked by a milking machine, with a vacuum pump and pipeline system.

The milk in the cow's udder flows directly from the udder, through the vacuum pipeline system to the cooling system, and from there to the main milk container.

We may hear a "persuasive explanation" from the kind farmers about how the vacuum in the pipeline milking system "pulls" (or "sucks") the milk from the udder to the main milk container.

Perhaps we should reflect on this explanation for a bit, even if it is said in a confident voice, by an experienced professional.

Does the vacuum in the pipeline system actually **"pull"** the milk out of the udder?

How is it possible that air pressure (even if it is low air pressure), like the pressure in the pipeline system, will cause the milk to flow **against it**?

A more correct explanation of the phenomenon shows that in the vacuum system, there is air and pressure. However, this is lower than the external air pressure.

The external atmospheric pressure is much higher than the pressure in the vacuum system (a whole atmosphere outside and only half an atmosphere in the vacuum system). The difference in pressures (about half an atmosphere) between the air pressure outside the udder and the air pressure in the vacuum pipeline is what causes the milk to flow from a high pressure to a lower pressure.

The high atmospheric pressure presses the udder **from the outside** and **pushes** the milk against the much lower pressure, which is in the vacuum pipeline system.

The pressure in the vacuum pipeline "tries" to push the milk in the opposite direction somewhat "unsuccessfully," if we are to judge by the results and by the quantity of milk in the main milk container...

Pushing or Pulling?

Two groups compete in "rope pulling." Each group holds the rope while facing the other group, and when the signal is given, the two teams start to pull the rope, each team pulling it to its own side.

When a handkerchief tied to the center of the rope passes beyond an agreed point (so that it will be on the side of the "stronger team"), this team will be declared the winner.

Are both teams actually **pulling** the rope?

If we place both teams in the same place that they were at the start of the competition, but this time we place them in such a way that the "pullers" face the outside, and have their backs to their opponents, and we keep the rules of the game as they were, we will clearly see that this time the teams are **pushing the rope**, and not **pulling** it, as we assumed previously.

Pulling

Pushing

Illustration No. 2

The way the force was exerted remains the same - the teams were **pushing** the rope previously as well, it is just that the way in which we were accustomed to think "whispered" to us that both teams were "pulling the rope."

Did the direction the competitors faced mislead us so much

that we replaced "pushing" with "pulling?" Perhaps we can see how much the concepts we have grown to become accustomed to, control the definitions with which we determine our approach to reality.

Perhaps we tend to replace pushing with pulling wherever we learn that there is an attraction force, and not as the result of a systematic observation of our surroundings.

A Tube and Balls

Let us imagine that we are in a closed room, with a straight, horizontal glass pipe crossing it throughout, entering through one wall and disappearing through the opposite wall.

In the room, not far from one of the walls, there is a device attached to the pipe, through which one can insert round balls into it, with a slightly smaller diameter than that of the internal diameter of the pipe.

After a ball is inserted into the pipe, it can move about freely throughout the pipe, if we give it a slight push.

In the first experiment, we insert a ball made of steel into the pipe, and to our surprise, we may see that the ball rolls through the pipe, without being pushed, towards the opposite wall, until it disappears from our sight behind the wall.

If we repeat the experiment once again, we will see that the ball rolls in the same direction as previously, and we may notice that its velocity accelerates as it approaches the opposite wall, until, with high velocity, it disappears through to the other side of the wall.

If we repeat this experiment several more times, we will see that in all the experiments, the results are the same.

What is happening here? Is there an explanation to the phenomena that we have seen so far?

We might be able to offer an explanation that may provide a suitable answer for all the results viewed so far:

"There must be a powerful magnet positioned on the other side of the wall, which pulls the steel balls towards it."

We are not able to confirm the assumption that there is a magnet behind the wall, since the room is closed, and we cannot leave it to examine this assumption.

We repeat the experiment once more, but this time, instead of using steel balls, we will use wooden or plastic balls, which are not "pulled" by a magnet: we expect the balls to stay still, in the same place they were placed by the device, since they were not pushed and the magnet has no effect on them.

To our surprise, the wooden balls will "behave" just like the steel balls. They will move with increasing speed towards the same wall, as if "something" behind the wall were "pulling" them towards it as well.

We repeat the experiment again, but this time we insert the balls into the pipe while pushing them towards the opposite wall, which is far from the assumed "source of pulling force."

Each ball we insert into the pipe will at first move quickly, according to the pushing force, but its velocity will decrease, until it stops and then starts moving back, with accelerated speed towards the "pulling wall," until it disappears from our sight, just as before.

It seems that our assumption about there being a powerful magnet behind the wall is incorrect, because the wooden or plastic balls, which are not pulled by a magnet, "behave" just like the steel balls.

Balls made of glass or any other material that is not affected by a magnet, will also "behave" in a similar way.

Could there be a stream of air flowing throughout the glass

pipe, **towards** the wall **to which** the balls rolled?

Air is **transparent**, just like the glass pipe, and thus, apparently, we do not observe its flow.

The ball which was inserted into the pipe was pushed by the strong air flow, since on the one side (where the air flow was coming from) there was a higher pressure than on the other side (towards which the air flowed), that is, there was a difference in air pressure on the two sides of the ball.

Thus, the ball is **pushed**, and its velocity accelerates until it reaches the speed of the air stream (we will not be able to see the ball reach its final speed, the speed of the air stream, as we are limited by the walls of the room in which the experiment is held).

When air is flowing through the pipe, the explanation of **being pushed** by air pressure is better than **being pulled** by a magnet (as we assumed earlier), as it suits all balls that were tried so far, regardless of the material of which they are made.

We repeat the experiment with the balls, but this time using balls made of wire mesh, so that some of the air flow can pass through the holes and escape to the other side of the ball. We will use two different kinds of balls for the experiment:

> ‣ Balls with large holes and thin wires: **light** balls.

> ‣ Balls with small holes and thick wires: **heavy** balls.

How will these wire mesh balls behave in the glass pipe, where there is an air stream flowing?

Some of us will expect the heavy balls to move faster, due to the large surface area of the wires which faces the air flow, and thus for a given pressure, a larger pushing force will be exerted upon them.

Others will expect the lighter balls to move faster, since a smaller force is needed to move them quickly.

In the experiment itself, we may see that the balls with the large holes and the thin wires change their speed **exactly** the same way as the balls with the small holes and the thick wires. (For clarity, we may ignore the air friction which occurs due to the movement of the balls).

A possible explanation for the fact that both types of balls move at the same acceleration and reach the same final velocity in the same period of time, is that although more force is needed to move the heavy balls, the large surface area of the wires creates a greater force on their side facing the air flow. This, in turn, will exert great force over them and move them faster. Balancing both phenomena will cause all the balls to move at the same acceleration and velocity.

This phenomenon reminds us of our early physics classes:

"All bodies fall (in an empty space) at the same velocity, or to be more precise **at the same acceleration,** that is: **the change of velocity in a unit of time."**

In the experiment we held with the balls, the air flow was meant to demonstrate the concept of "pushing" versus the concept of "pulling" and to enable us to choose between them again.

Perhaps the explanation given to bodies getting closer to earth due to "an attraction force" is not the only possible one, and perhaps it is also not "the best one," as it does not let us understand the observed phenomena well.

It is possible that doubting the explanations that most people agree upon regarding "the cause" of these phenomena could improve our way of thinking.

The doubts mostly regard the **mechanism** which causes bodies to fall or to draw closer to each other.

What causes a body to be "heavy"?

What causes the phenomenon that we call "attraction"?

Could it be that the reason why up to today, there still has not been a satisfying physical explanation for the way in which the attraction force works, is that **there is no attraction force?**

Could it be that there is a simple explanation, which is different from what is commonly thought, that could enable us to both understand and to unite several fields of science, which we cannot unify while we continue to adhere to the theory of "attraction force?"

If it turns out that "attraction force" contradicts one of Newton's basic principles, the principle of inertia, we might ask, what concept is better: **attraction force** or **inertia?**

ATTRACTION

*"When Newton demonstrated that the force which acts on
each of the heavenly bodies depends on its relative position
with respect to the other bodies, the new theory met with
violent opposition from the advanced philosophers of the day,
who described the doctrine of gravitation as a return to the
exploded method of explaining everything by occult causes,
attractive, and the like. Newton himself, with that wise
moderation which is characteristic of all his speculations,
answered that he made no pretense of explaining the
mechanism by which the heavenly bodies act on each other.
To determine the mode in which their mutual action depends
on their relative position was a great step in science, and
this step Newton asserted that he had made. To explain the
process by which this action is effected was a quite distinct
step, and this step Newton, in his Principia, does not attempt
to make."*

JAMES CLERK MAXWELL

In order to examine the attraction force in its physical sense, we
should temporarily ignore all the knowledge that we have gained
up to today and re-examine the concept of "force" as a physical
reality and not as a conceptual model.

The theoretical experiment we shall do is to examine an
assumed way in which force works, while assuming that the force
belongs to the world of particles or waves and motion, and is not
an abstract, vague, or even somewhat mystical concept.

Our starting point will be the **mass**, which is considered to be the source of the thing called "attraction force." We will try to build a theory that may explain, in terms of particles and motions, how the attraction force works. For this purpose, we shall assume that in the whole universe, there are only two bodies, or two singular masses, that we shall term, as usually accepted in school books: mass **A** and mass **B**.

Nothing else exists in the whole universe that we created, apart from these two singular masses.

According to the accepted attraction theory of gravity, mass **A** will pull mass **B**, and equally so mass **B** will pull mass **A** - and as a result, they will both move towards each other, until they collide into each other.

In order to build a tangible theory to explain this phenomenon, if it would, indeed, occur, we will make another assumption:

"Attraction force works through particles, waves, or any other real "thing," even if it has not yet been discovered."

At the first stage, we will focus our attention on mass **A** pulling mass **B** towards it, and we will assume that this "real thing" that derives from mass **A**, exits mass **A** and moves from mass **A** towards mass **B**.

We will name that "thing" **UWP**, an acronym for: **U**ndiscovered **W**aves or **P**articles.

The UWP leaves mass **A** and moves towards mass **B**, and according to the accepted theory of attraction, it will cause mass **B** to move towards mass **A**, that is, to move **in the opposite direction** to its original direction (the same thing occurs, of course, to the UWP of mass **B**).

Could it be that the UWP that left mass **A**, and moved towards mass **B**, suddenly changed its direction and caused mass

B to move towards mass **A**, that is, against, or opposite to its previous direction?

Had the UWP **not** changed its direction, we would have seen many UWPs moving from mass **A** towards mass **B** (and in other directions too), and mass **B** would have moved towards mass **A**, **against** the flood of UWPs coming towards it.

Had the UWP not changed its direction, it would have directly hit mass **B** and caused it to move away from mass **A**, that is, in the same direction as that of the UWP.

Does the principle of inertia not apply to the UWP of mass **A**?

Why did it change its direction?

If the UWP did not change its direction, how did mass **B** move against the direction of the stream of UWPs, towards mass **A** and not in the opposite direction, namely, away from mass **A**?

To this day, no element has been observed to **leave** one mass, move towards another mass and make the second mass move **towards** the first mass.

If a particle such as the UWP had been discovered, it would have defied the principle of inertia, and would have "forced" us to change our way of thinking and our understanding of physics significantly.

If we accept the UWP as "the carrier of attraction," we should probably have to deduce that, **attraction and inertia contradict each other, and cannot co-exist in the same physics.**

In a physical theory composed of a limited number of basic assumptions and built solely on the elements of matter and motion, a UWP that leaves one mass and moves towards another mass will not defy the principle of inertia, as long as it continues to move with a constant speed in a straight line.

By doing so, the UWP would **push** any object that is in its way; however, by changing its direction and making a material body move in an opposite direction to its previous direction, at that very moment it **"expelled itself from physics as we know it,"** and we should probably have to phrase new basic principles.

Does the UWP not remind us of the "bosons," "gravitons" or "gluons" that many physicists consider to be the "carriers of the force" among the elementary particles of the atom, according to the model termed the "Standard Model of the Atom"?

Do the UWPs, the "bosons," "gravitons" and "gluons" have a right to exist in a world where inertia exists?

If we consider inertia to be a basic principle in nature, should we accept a growing number of exceptions to the principle, or would it be better to get rid of them?

Is it possible that the laws of nature are not really laws, and do not come from nature, but are just certain interpretations that we humans give to the phenomena or the events that we observe?

Perhaps we should see "the laws of nature" as a wonderful invention of the spirit of man, as he tries to understand his experiences through his senses, rather than as laws that exist in the universe which were "revealed" to us in some way.

Bodies approaching each other or bodies falling towards each other are **observed facts**, acquired through our senses.

The **attraction force** explains the observations, but is **not a fact, and is not a law of nature.**

Perhaps the attraction theory does not advance our thinking today or suggest additional experiments to verify or refute it, or unite it with other theories.

> ➤ Perhaps the concept of "attraction force," when we express it as the motion of particles or waves, and which we derive through the observation of "pulling bodies," **contradicts the principle of inertia.**

> ➤ The repelling or pushing of one body by another body in

contrast to what we might consider as being an attraction force, does not contradict the principle of inertia, or any of the other laws of physics.

> Perhaps attraction cannot exist between two masses in a world where there are only two singular masses, with no other masses around them.

> These masses will never approach each other, but will repel each other and may drift away to the edges of an empty space **without contradicting the principle of inertia.**

If we add more masses to a universe that has only two singular masses, would the behavior of the masses change and would its laws change?

Chapter 2

The Fable of the Middle East Tree Frog

> *"Owing to our senses' feebleness, we are not able to determine the truth."*

Anaxagoras

Cute, tiny frogs, which are greenish-grey, no bigger than 4 cm long and weigh no more than 7 grams, belong to a species called the Middle East tree frog.

We found this description in one of the old books:

"The Middle East tree frog jumps well, as its back legs [which] are longer and stronger than its front legs, help it jump from branch to branch."

With the help of these Middle East tree frogs, we will examine the "attraction theory" that is based only on pushing, and is presented here as a fable or a tale.

The tale starts with a mesh ball, placed in space and unaffected by any attraction or pushing forces from any direction. Around it, at a set distance from it, are thousands of tiny air openings, which are at an equal distance from each other.

Middle East tree frogs live on the ball very happily, and will, at times, jump cheerfully to the treetops of the small, bare trees, which grow on the tiny net ball. In order for them not to pass by the treetops and disappear in the dark space (as mentioned above, there is no force of attraction on the ball), the Middle East

tree frogs hold onto the branches with the little sticky pads at the tips of their toes.

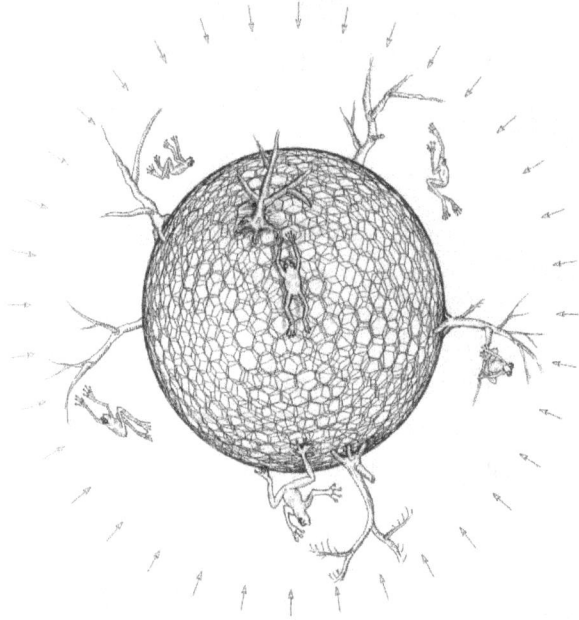

Illustration No. 3

Thus the tale continues:

"One day, jets of air burst came out from all the openings around the mesh ball, towards that 'planet'..."

The jets of air which reached the tree frogs from their backs, directly from the air openings above them, pinned them down to the mesh ball, while the jets of air that came from the other side of the ball and passed through the mesh, hurt their stomachs and "tried" to keep them away.

Some of the air molecules from the other side "did not manage" to pass through the mesh wires which filled the inside of the ball, which is why the jets of air that came "from the bottom" were slightly weaker than those which arrived "from the top."

The forces exerted by the jets of air over the tree frogs in every direction but "upwards" or "downwards" were equal and each force neutralized its respective counterpart.

"When the tree frogs tried to jump onto the treetops, as they were used to doing from an early age, they were lifted above the mesh at a high speed. This gradually decreased, since the air pressure on their backs was greater than the air pressure exerted on their stomachs, so that at the end, their speed, as they went upwards, reached zero and then they fell again onto the face of the tiny mesh planet, at an increasing speed. How embarrassing…

And since then, the Middle East tree frogs fall back to the mesh ball, every time they try to skip to the sky."

That is the tale.

The tale of the tree frogs can raise a few questions:

> Does the description of the mesh ball match what we know about the structure of matter?

> Are masses dense, impenetrable concentrations of matter, or perhaps they may be seen as a net which has small "cores of matter" scattered in defined volume in space, relatively distant from each other?

> Is it possible that the jets of air that reached the tree frogs from all directions can open to us new directions for thinking and experimenting, beyond what is currently known?

In terms of "gravitation," is there any resemblance between the condition of the human species and other species on earth, and the grim condition of the tree frogs that jumped on the surface area of the tiny mesh ball?

FRACTICLES AND REPULSION

In recent decades an unfortunately noticeable lag has taken place between theory vs experimentation ... It is possible that substantive progress will be achieved but only through ideas involving more radical concessions with regard to concepts that theoretical physicists have become accustomed.

SHMUEL SAMBURSKY

The distances between the atom's nucleus and the electrons that surround it, are, as far as we know, relatively larger than the distances between the sun and the earth and the rest of the planets in the solar system.

Most of the volume of the solar system, from the sun to the furthest planet that we know of today, Pluto, is an **empty space**, with just a minuscule quantity of matter per unit of volume (maybe a single atom of hydrogen for every cubic km of space).

Most of the volume of the atom is also empty. Only a tiny part of it is "occupied" by the nucleus and the electrons that move around it.

The atom nucleus is, as far as we know, composed of different particles. The mass of these particles takes up at most, a minuscule part of the volume of the atomic nucleus. A similar situation may exist within the electrons themselves, which encircle the nucleus from afar.

Since most of the matter we know about is composed of

atoms, whose volume is mostly empty space, with empty space between them, we can assume that very small particles that move very fast will be able to pass **through** the matter or the mass from one side of the earth, for example, towards its other side, without stumbling upon any kind of "material nucleus."

Let us assume that every sub-atomic particle occasionally emits a "fracticle" many orders of magnitude smaller than it, which will continue its movement in a straight line at a constant speed of light, as long as it is not stopped or diverted from its route by another sub-atomic particle or another fracticle in its way.

The term "fracticle" distinguishes between the known, sub-atomic "elementary particles" (protons, neutrons, electrons and so on), which we assume exist, and other particles, which are several orders of magnitude smaller than them and have not yet been discovered.

"Fracticles" form the basis of the theory of **repulsion** which we present here, instead of the commonly accepted **attraction** theory.

The Assumed Properties of Fracticles

When a fracticle hits one of the sub-atomic particles in the atomic nucleus or in the electrons that move around it, it "disappears" or gets swallowed up into that particle. This causes that particle to deviate slightly from its course into the direction of the previous movement of that fracticle (some fracticles may also be diverted and not get swallowed up).

A significant number of fracticles may disappear within a sub-atomic elementary particle, before it emits another fracticle. In other words, the rate of "swallowing" fracticles by each sub-atomic particle is faster than the rate of emitting fracticles, and therefore, it is possible that the mass of atoms in the universe is growing.

Since the atoms of measuring equipment are also "growing" relatively, it is possible that the phenomenon is unobservable.

Perhaps we can say that the momentum of the fracticles is conserved, but their mass is not conserved as understood according to the law of conservation of energy.[11]

If fluxes of fracticles coming from different directions were to hit the atomic nucleus, the atomic nucleus and the whole atom would move in the direction of the numerically biggest flux.

When a single fracticle collides with an electron, it is diverted from its course, according to the angle of the hit, and the electron will also change its movement according to the mass relationships between them and its direction of movement before the collision.

If the sub-atomic, elementary particles emit fast little fracticles, a huge flood of fracticles will come to earth (and to other masses) from all directions, from all the masses in the surrounding universe.

If indeed fracticles are several orders of magnitude smaller than any particle discovered so far, they would probably be able to pass between the atoms and between the particles that compose the atoms, without hitting any material body.

A Universal Flux of Fracticles

We may term the large number of fracticles which come to us from every direction a "universal flux of fracticles." The flux can

---

11   There are other options apart from rejecting the Law of Conservation of Energy in its current form. Since this problem is fundamental to the core of Fracticle Theory, additional research is needed to develop and complete this subject. The closest particle to fracticles is the neutrino, which has already been discovered (The author).

be measured quantitatively according to the number of fracticles that pass through a given cross-section in a unit of time.

We can imagine that these miniscule fracticles, which arrive from all around, may pass through earth without colliding into any kind of "matter," as if earth is **almost** transparent to them.

The self-emission of fracticles by earth or any other singular mass is **negligent** in its quantity as compared with the flux of fracticles arriving from every direction from every mass in the universe.

Not all fracticles will "manage" to pass in their course through the "matter" of earth, without colliding into the particles which build the atoms from which earth is composed. Some of them will directly hit sub-atomic particles, which will halt the movement of those fracticles.

Due to these random collisions, fewer fracticles will pass through earth than the total number of fracticles which reached its surface from all the masses in the universe.

At each point on earth, the flux of fracticles coming "from above" (from outside earth) will be numerically larger than the flood of fracticles passing through it and coming "from below" (as mentioned earlier, this flux is partially "blocked" by all of the sub-atomic particles from which the earth is composed).

The excess pushing force exerted by the fracticles coming "from above" onto the masses on the ground, as compared with the pushing force of the fracticles coming "from below" to the crust of the earth, is due to the **large number** of fracticles coming "from above," which directly hit the masses that are in their way, versus the relatively small number of fracticles which "manage" to pass through earth and reach its crust and the masses from the underside.

The number of fracticles swallowed up by masses is small, in relation to the flux of fracticles coming from all of the galaxies, from the planets and the many sub-atomic particles that fill the whole of interstellar and intergalactic space.

The number of fracticles which hit the earth (or other masses) from every direction without having any obstacles in their way is more or less equal from all directions, and thus the forces they exert over these masses are balanced.

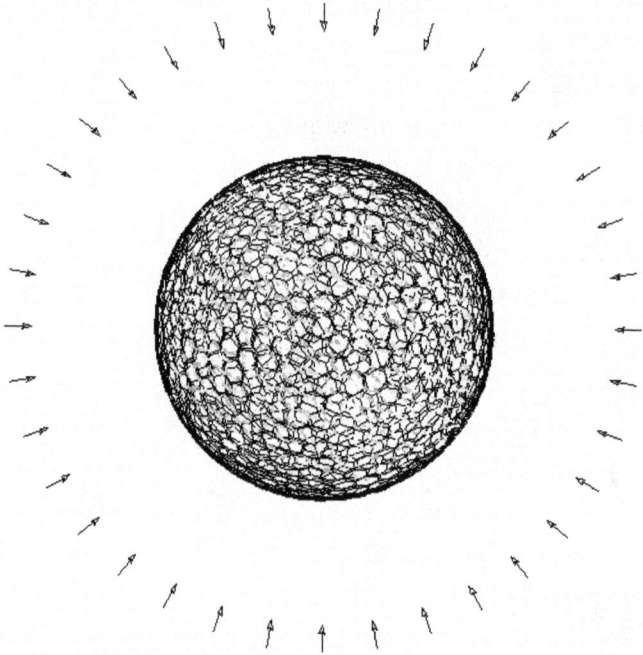

Illustration No. 4

The Action of the Fracticles

The excess pushing force exerted by the fracticles coming "from above" as they collide into bodies they hit, as compared with the pushing force exerted by the fracticles coming "from below" (which manage to penetrate the mass – earth), is commonly termed as "**attraction**," "**gravitation**" or "**curved space-time**."

This force is greater towards the center of the earth, since the number of fracticles blocked by the mass of the earth in this direction is greater than the number of fracticles blocked by its mass as they pass shorter routes (not through the center of the earth).

We use the term "**mass**" to describe the property of a body to halt or block the flood of fracticles that pass through it to some extent, so that there is **a difference** between the number of fracticles passing through it and those which come from the opposite direction.

Mass

According to the repulsion theory, one can define mass as "the quantity of fracticles that are blocked by a body in a unit of time."

One may see the mass as the main cause of the numerical difference between two (or more) similar floods of fracticles coming from opposite directions, when the measurement takes place in a certain, defined cross-section of the mass.

> It is possible that the phenomenon "matter," which is detected by our senses (or by the equipment which we created to enhance them) is "a numerical imbalance" in the flood of fracticles which comes to us from all directions.

> The **blocking** of the flood of fracticles by a material body is commonly known as "**attraction**," or, according to the theory presented here, as "**repulsion**."

> As the numbers of atoms in a body, or the sub-atomic particles in a body, are greater, this body will block a greater part of the flood of fracticles passing through it. That is why the force of repulsion **from the outside turning inwards** will be greater.

> ‣ A body with high density contains more sub-atomic particles per unit volume than a body with low density. Thus, it will block more fracticles and the repulsion to its surface will be stronger.

> ‣ A body which does not cause a disruption in the flux of fracticles and does not create a numerical difference in the flux of fracticles will be "transparent" to the flux of fracticles, with no mass, and is undetectable by us.

To sum up this chapter, let us imagine two bodies which have **identical density**, but the volume of one is bigger than the volume of the other, that is, its mass is greater:

This body will block more fracticles from the cosmic flux and the repulsion to its surface will be stronger, or in more accepted language:

Its "attraction" will be greater than the "attraction" of the body with the smaller volume.

The sum of the fracticle "vectors" created by the presence of mass causes their mutual distancing or coming together.

One may also say that "gravity" is created by mass, but is not a property that derives from the mass, but a phenomenon created by an element which reaches the mass externally.

Considerations in Choosing a Theory

"A theory which is not refutable by any conceivable event is non-scientific."

Karl Popper

In H. G. Wells' book, *The* First Men *in the* Moon, Dr. Cavor, one of the protagonists in the book, develops a new material, which can negate the force of gravity.

The magic material is named Cavorite, after its inventor, Dr. Cavor.

In the book, a body that is situated above the Cavorite becomes weightless and earth does not have any effect over it that could 'attract it' to the ground.

Let us follow a strange accident that happened during the preparation of the Cavorite...

But Cavor's fears were groundless, so far as the actual making was concerned. On the 14th of October, 1899, this incredible substance was made!

Oddly enough, it was made at last by accident, when Mr. Cavor least expected it.

He had fused together a number of metals and certain other things--I wish I knew the particulars now!--and he intended to leave the mixture a week and then allow it to cool slowly...

But it chanced that, unknown to Cavor, dissension had arisen about the furnace tending. Gibbs, who had previously seen to this, had suddenly attempted to shift it to the man who had been a gardener, on the score that coal was soil, being dug, and therefore could not possibly fall within the province of a joiner; the man who had been a jobbing gardener alleged, however, that coal was a metallic or ore-like substance, let alone that he was cook.

But Spargus insisted on Gibbs doing the coaling, seeing that he was a joiner and that coal is notoriously fossil wood. Consequently Gibbs ceased to replenish the furnace, and no one else did so...

And then--

The chimneys jerked heavenward, smashing into a string of bricks as they rose, and the roof and a miscellany of furniture followed...

"...It was not an explosion. It's perfectly simple. Only, as I say, I'm apt to overlook these little things. Inadvertently I made this substance of mine, this Cavorite, in a thin, wide sheet... You are quite clear that the stuff is opaque to gravitation, which it cuts off things from gravitating towards each other?"

"Yes," said I. "Yes."

"Well, so soon as it reached a temperature of 60 degrees Fahrenheit, and the process of its manufacture was complete, the air above it, the portions of roof and ceiling and floor above it ceased to have weight. I suppose you know--everybody knows nowadays--that, as a usual thing, the air has weight, that it presses on everything at the surface of the earth, presses in all directions, with a pressure of fourteen and a half pounds to the square inch?"

"I know that," said I. "Go on."

"I know that too," he remarked. "Only this shows you how useless knowledge is unless you apply it. You see, over our

Cavorite this ceased to be the case, the air there ceased to exert any pressure, and the air round it and not over the Cavorite was exerting a pressure of fourteen pounds and a half to the square in upon this suddenly weightless air. Ah! You begin to see! The air all about the Cavorite crushed in upon the air above it with irresistible force. The air above the Cavorite was forced upward violently, the air that rushed in to replace it immediately lost weight, ceased to exert any pressure, followed suit, blew the ceiling through and the roof off....

"You perceive," he said, "it formed a sort of atmospheric fountain, a kind of chimney in the atmosphere. And if the Cavorite itself hadn't been loose and so got sucked up the chimney, does it occur to you what would have happened?"

I thought. "I suppose," I said, "the air would be rushing up and up over that infernal piece of stuff now."

"Precisely," he said. "A huge fountain--"

"Spouting into space! Good heavens! Why, it would have squirted all the atmosphere of the earth away! It would have robbed the world of air! It would have been the death of all mankind! That little lump of stuff!"

"Not exactly into space," said Cavor, "but as bad-- practically. It would have whipped the air off the world as one peels a banana, and flung it thousands of miles."

If Cavorite were available to us, we could empirically examine which theory offers a "better" explanation to more observable facts: that which sees in the masses approaching each other a result of an attraction force derived from the masses themselves, or that which explains the same facts with repulsion derived from a flux of fracticles coming from outside those masses.

➤ If we assume that the Cavorite is equally not penetrable to attraction or repulsion, we can conduct a simple intellectual experiment. In this experiment, we can place precise electronic scales on a Cavorite board and we place a test weight on the scales of, say, 1 kg, and observe the results of the weighing.

According to the attraction theory:

If the force of attraction cannot pass through the Cavorite board, on which we placed the accurate scales and the weight, the weight will weigh much less than 1 kg, and its heaviness might even disappear completely (depending on the Cavorite's efficiency).

According to the repulsion theory:

On the other hand, if we were to accept the repulsion theory with the assumption that the fracticles cannot pass through the Cavorite board, the undisturbed flux of fracticles coming "from above" would have pushed the weight "downwards" to the precise scales. At the same time, the flux of fracticles coming "from below" is mostly (or completely) stopped by the impenetrable Cavorite board, and does not reach the weight to push it upwards.

The difference between the flux of fracticles coming from above and the one coming from below was bigger than the "usual" difference, and would have caused the weight to be pressed down with great force, which would have exerted much greater force on the scales than 1 kg (maybe 2 kg).

➤ If we were to place the Cavorite board above the scales and the weight,

According to the attraction theory:

The weight would be exactly 1 kg according to the attraction force. This is since the attraction force, which is derived from earth, comes from below to the weight, without the interference or effect of the Cavorite board.

According to the repulsion theory:

The weight would become lighter than 1 kg, since the flux of fracticles coming "from above" would mostly be halted by the Cavorite board, without pushing the weight "downwards." At the same time, the flux of fracticles coming "from below," which were unaffected by the Cavorite board, would continue to push the weight "as usual" - "upwards." The difference between the repulsion forces will cause the weight to "appear" to be light! (Or zero).

If such an experiment were possible, it would offer a clear and conclusive answer regarding which of the two theories provides a better explanation to the observable facts, and would make our decision easier.

We can reach interesting conclusions from the "Cavorite experiment:"

A. If we were to place scales on top of a Cavorite board, and on it we would place a weight of 1 kg, and the scales would show less than 1 kg, the conclusion would be: **there is attraction**.

B. If we were to place scales on top of a Cavorite board, and on it we would place a weight of 1 kg, and the scales would show more than 1 kg, the conclusion would be: **there is repulsion**.

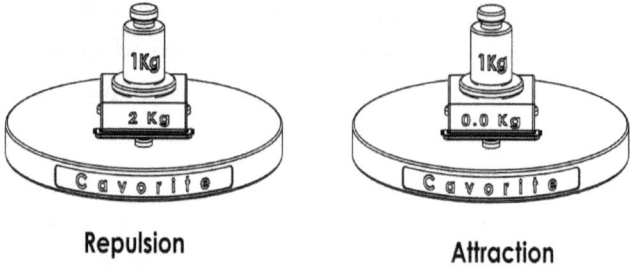

Repulsion   Attraction

Illustration No. 5

C.   If we were to place a Cavorite board **above** a scale, and a weight of 1 kg, and the scale would show exactly 1 kg, the conclusion would be that **there is attraction.**

D.   If we were to place a Cavorite board **above** a scale and a weight of 1 kg, and the scales would show less than 1 kg, the conclusion would be that **there is repulsion.**

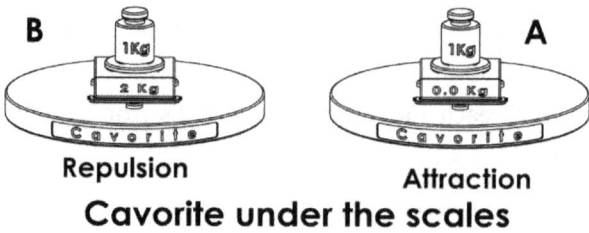

B   A

Repulsion   Attraction

Cavorite under the scales

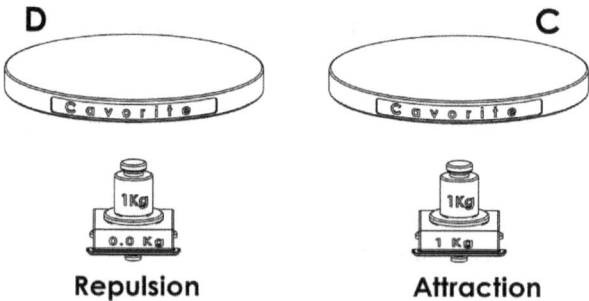

D   C

Repulsion   Attraction

Cavorite Above the scales

Illustration No. 6

Can one conduct a similar experiment without any "Cavorite?"

What should we expect, if we take a mesh board in the shape of a thick disc, and place accurate scales beneath it, on which there is also a weight of 1 kg, and we "shower" the disc with a "rain of peas?"

> ➤ The rain of peas will pass through the mesh disc almost undisturbed, will hit the weight and cause the scales to show a weight that is more than 1 kg (1 kg of weight + the pressure of the rain of peas hitting the weight).

> ➤ If we turn the round mesh disc around its center, or back and forth, parallel to the disc level, we might manage to see that the rain of peas passing through it is diminishing, since more peas are hitting the mesh wires, as they pass through the thickness of the disc. These peas are forced to the sides or bounce among the mesh wires, but they do not manage to pass in their original direction and speed.

> ➤ If we increase the rotation speed of the mesh disc, we may be able to see that the number of peas passing through it is smaller and the weight of the weight which we placed on the precise scales is decreasing, since the rain of peas pushing it towards the scales is weakening.

> ➤ The relation between the weight of the weight on the scales and the rotation speed of the disc, depends on the extent that the mesh disc disrupted the free passage of the peas.

One might also think about another option, where the mesh disc revolving around its axle caused the **weight to be drawn** towards it (Attraction) and so made it **seem lighter**.

> ➤ Had we not seen the "rain of peas" with our own eyes, we would probably have chosen the "attraction" option. However, after observing the rain of peas and the increase in the weight caused by it, we would, of course, prefer the repulsion diminishing as the cause for reducing the weight in the fast rotation of the mesh disc blocking the rain of peas.

> ➤ Since we are unable to see the small fracticles, which are several orders of magnitude smaller than any "known" sub-atomic particle, unlike with the visible "rain of peas," the choice between the theories may be a difficult one.

What should we expect, if instead of a mesh disc, we take a metal disc, as dense as possible, such as a thick platinum disc of 21.45 g/cm^3 density, and place this disc above the scales and the test weight, similar to the imaginary experiment with the Cavorite?

Before placing the platinum disc above the scales and the weight, our accurate scales showed exactly 1 kg, just as was written on the weight.

After placing the platinum disc above the accurate scale and the weight, they will show that the weight weighs slightly less, due to the "attraction" which the platinum mass is exerting over it according to the accepted attraction theory, or due to blocking the fracticles coming from above, according to the proposed repulsion theory.

Now, we will turn the platinum disc around its axle at an increasing speed and we will continuously check how much the weight weighs, as shown by the accurate electronic scales:

> ➤ According to Newton's attraction theory, the weight will still weigh the same (1 kg), since the attraction comes "from below," from the center of the earth, and does not pass through the rotating platinum disc.

> According to the repulsion theory (as we saw in the "rain of peas" experiment), the weight will weigh less, since the flux of the fracticles coming from above is "disrupted" from its course, and the flux of the fracticles coming "from below" remains unchanged.

Repulsion and the Theory of Relativity

According to Einstein's theory of relativity, it is commonly assumed that the weight on the scales will be lighter, since every mass that is in motion will have its mass increased.

> The larger mass of the spinning platinum disc (due to its motion) will cause the weight to be "attracted" to it, and the scales on which it is placed will show a lower value than the one marked on the weight.

> If the speed of the platinum disc reaches the speed of light, it will have an infinite mass, which will pull the weight placed beneath it upwards with an infinite force.

> According to the theory of relativity, accelerating the rotation of the platinum disc creates a larger attraction force around it than the attraction force of that disc while at rest. This is contrary to Newton's attraction theory, but according with the repulsion theory through the influence of fracticles.

> The faster a body spins, the fewer are the fracticles that will pass through it, and the larger will be the difference between the forces pushing every other body toward it. Thus, if the speed of any part of the disc reaches the speed of light, fracticles whose speed is smaller or equal to the speed of light will not be able to pass through it, and the force which pushes other bodies towards it will reach its ultimate maximum in our universe.

> Using terminology from the theory of relativity, this part of the disc will have an "infinite mass" and the attraction forces towards it will be "infinite."

> We are probably unable to imagine an infinite mass or an infinite force, and this concept has no clear meaning to us. Therefore, we will suffice with the **tendency** and not the final result.

> According to the repulsion theory, a body which is near another body spinning at the speed of light, will be pushed towards this body, with great force, that equals all the force exerted by the flux of fracticles coming from the opposite direction to the spinning body (for example, from the center of the earth), with no balancing counter force.

> If the universe in which we live is "a finite universe," then the force exerted by the flux of fracticles coming from a specific direction (from all masses in the same direction) will be great but finite.

> It is difficult for us to imagine the size of such a force, as it has no counterpart in our world of concepts and we also do not know the density of the flux of fracticles. Yet, since the flux of fracticles coming from a finite universe is also finite, however great its force may be, it has some finite value.

> If the universe is "infinite," the density of the flux of fracticles coming to us from all directions will be infinite and the forces exerted will be infinite forces.

> This possibility does not seem probable, since an infinite flux cannot pass instantaneously from all directions. If there is a fracticle at each and every point in space, the fracticles will not be able to move from place to place, as every place in the universe will already be "taken" by another fracticle.

We do not need to spin the disc at the speed of light or in accordance with any definition that includes the concept of "the infinite," since any significant acceleration of the spin of the platinum disc will probably provide us with useful conclusions.

Following the intellectual experiment with the spinning disc, we can conclude that even if we have no Cavorite, we can see that the repulsion theory is in line with the theory of relativity (to a point), but it contradicts the explanation of the attraction force in its classical sense, as it was understood after the theory was formulated by Sir Isaac Newton.

If we accept the repulsion theory as preferable to the attraction theory, we can draw some interesting conclusions:

> ➤ If a body rotating at the speed of light does not let fracticles pass through it, then the speed of the fracticles that create the repulsion is at most the speed of light (which is also the maximal speed in the universe, according to the general theory of relativity).

> ➤ If the speed of the fracticles that create the repulsion force is the speed of light, then the effect of the repulsion, or "the gravitation," occurs at the speed of light and is not immediate, as predicted from Newton's "classical" theory (in his gravity formulations, neither time nor speed are mentioned).

> ➤ If indeed "gravity" spreads at the speed of light, we might be able to examine this assumption through some experiment. Such an experiment could reinforce the repulsion theory, until a better explanation is found.

> ➤ It is possible that light, and what is termed "electromagnetic waves," are a certain part of the "universal flux of fracticles," which create what is termed "gravity."

> ➤ If a "cosmic accident" were to happen, where the sun would evaporate **immediately**, without leaving any material trace in our world, then we, earth and humanity, would probably keep rotating around the place where the sun used to be for about 8 minutes more. This is the time required for sunlight and gravity to reach earth, and the time before earth would go on its "final way" in a straight line and at a steady pace, into the dark depths of space, along with humanity's temporary remains.

Although their fates would be sealed, it is likely that the last members of the human race would kill each other, either because they do not belong to the same nation or faith or because each person has a different view or ideals that are "supreme" compared to that of his peers.

If we adopt the repulsion theory, because it offers better qualitative explanations to observable phenomena, we will have to examine whether we can also receive quantitative answers, as provided by the classical attraction theory.

If we are able to obtain a quantitative answer, which matches the qualitative explanation for the repulsion theory, we will be able to see it as another step towards preferring this theory.

Can one examine the effect of repulsion quantitatively?

DISTANCE AND REPULSION FORCE

"The notion of the gravity force, with those who admit Newton's law, but go with him no further, that matter attracts matter with a strength, which is inversely as the square of the distance."

MICHAEL FARADAY, "EXPERIMENTAL RESEARCHES IN ELECTRICITY"

In order to examine whether the repulsion theory offers us quantitative answers to observed phenomena, rather than just qualitative answers that explain the relation between the phenomena we receive (through our senses or our equipment), we will conduct another intellectual experiment, assuming the fracticles theory.

Let us imagine a thin rod, which is one meter long and has a diameter of a single atom.

As is the custom, we will mark the ends of the rod with the letter **A** and **B**, and we will call the length of the rod **A-B**.

We will place a small material sphere, which we will term C (Illustration No. 7), at the height of **H**, above the center of the rod **A-B**, and we will ask:

What will be the "repulsion force" that will be exerted over point **C**, in the direction towards the rod **A-B?**

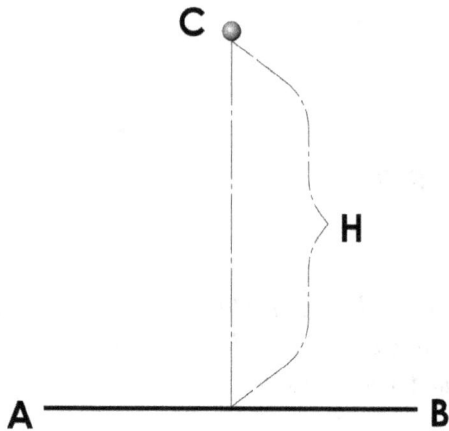

Illustration No. 7

Had we not placed the rod A-B next to point C, then the fracticles reaching C from all directions, from across the universe surrounding point C, would have created a "ball of repulsion," in the center of which would be C, as described schematically in Illustration No. 8 (below).

If we assume that the density of the fracticles arriving from all directions is more or less equal, then the repulsion forces exerted on C will be counterbalanced, since in front of every fracticle coming from a certain direction, a similar fracticle will be coming from the opposite direction.

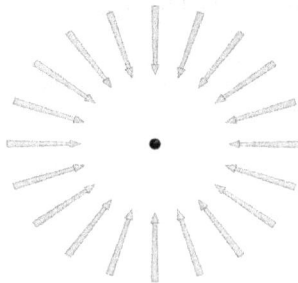

Fracticles coming from all directions
Illustration No. 8

The pushing force created by each fracticle hitting C from one direction is counterbalanced by the collision of a similar fracticle coming from the opposite direction.

One may say that the sum of the repulsion forces exerted on point C equal zero, since every push by a fracticle is counterbalanced by a push from the opposite direction.

By placing the rod **A-B** at a distance of **H** from point **C**, part of the flux of fracticles was blocked by rod **A-B**, and fewer fracticles will reach point **C** from the direction of rod A-B (as compared with the other directions).

One can present the area where fewer fracticles will move about as a triangle with the vertices A-C-B, where the density of the flux will be lower than in other directions around vertex C, as described in Illustration No. 9:

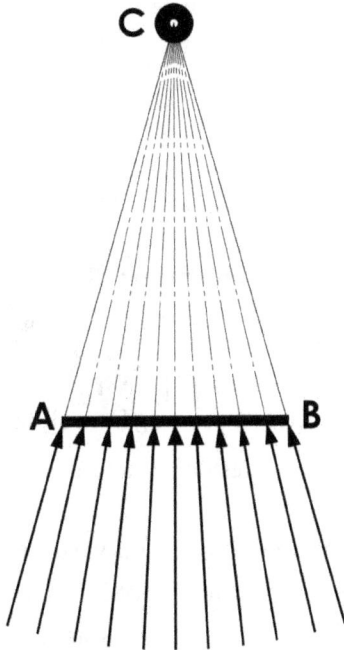

Illustration No. 9

We will term triangle **A-C-B,** *The Shadow Triangle of C.*

We will term the triangle that opposes the Shadow Triangle, and is created on the other side of vertex C at the distance of H' from vertex C, Triangle A'-C-B', and name it *The Repulsion Triangle.*

The Repulsion Triangle includes all the fracticles arriving to vertex C from the direction opposite the Shadow Triangle, as described schematically in Illustration No. 10 below:

The distance **H,** from vertex C to rod A-B, is equal to distance **H'** from vertex **C** to the base of the Repulsion Triangle **A'-B'.**

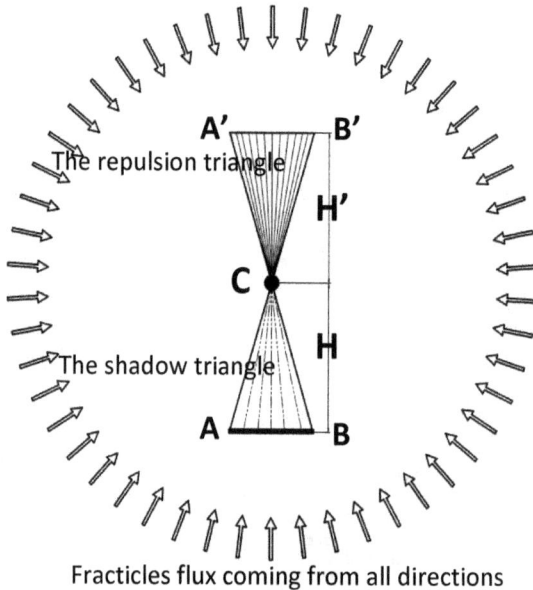

Fracticles flux coming from all directions

Illustration No. 10

The repulsion forces exerted on vortex **C** towards rod **A-B** equal the difference in the number of fracticles crossing side **A-B** at a particular moment and moving towards vortex **C,** in contrast to the (larger) number of fracticles crossing the imaginary side

A'-B', and reaching vortex **C** from the opposite direction, without stopping by rod **A-B**.

If we shorten the length of rod **A-B** by half, the base of the Shadow Triangle will be shortened by half, and to the same extent, the base of the Repulsion Triangle will be shortened by half, and with it, the difference in the number of fracticles arriving at vortex **C** in a unit time. Thus, the repulsion force toward vortex **C** from rod **A-B** will be reduced to half.

If we raise vortex **C** to double the height over rod **A-B**, so that its height will be 2 **H**, the new Shadow Triangle **A-C-B** will be twice as high as the previous Shadow Triangle (Illustration No. 11).

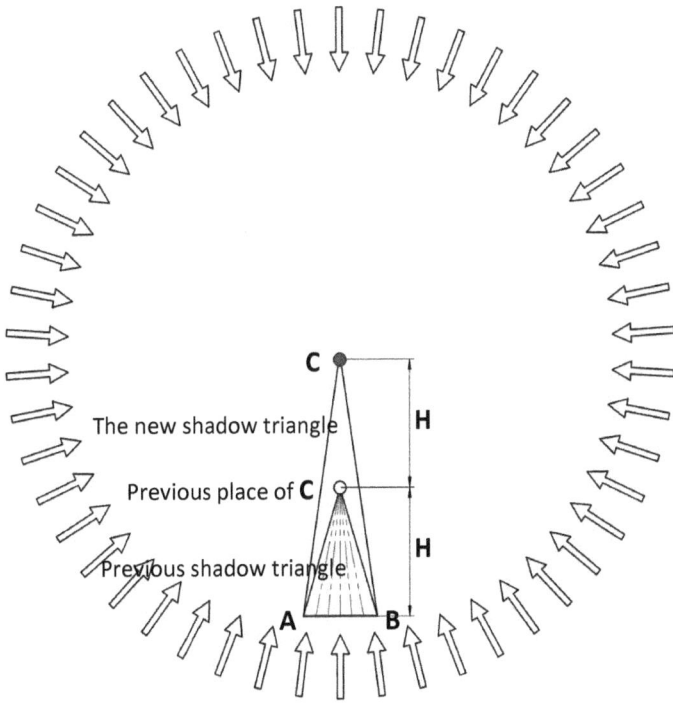

C

H

The new shadow triangle

Previous place of **C**

H

Previous shadow triangle

A   **B**

Fracticles flux coming from all directions

Illustration No. 11

If we draw a parallel line to rod **A-B** where vortex **C** was previously, and we mark the points of intersection of this parallel line in the Shadow Triangle with the letters **D** and **E**, we will see that triangles A-C-B and E-C-D are "similar triangles," according to three angles, as described in Illustration No. 12.

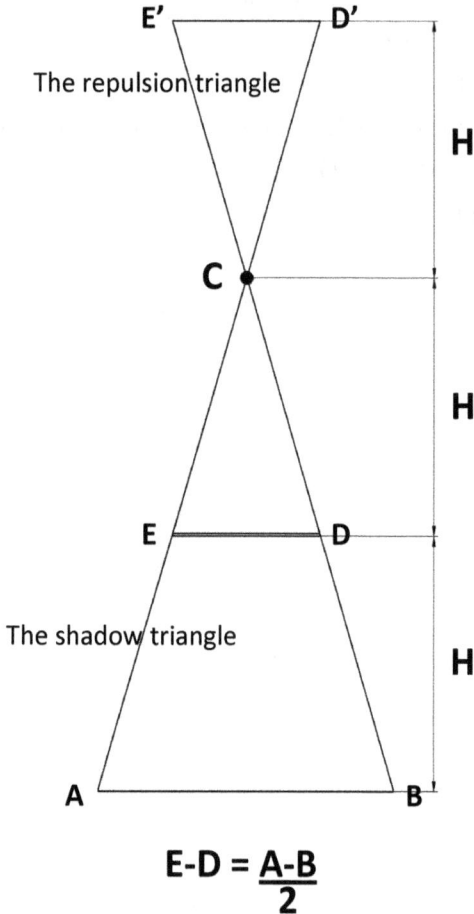

$$E\text{-}D = \frac{A\text{-}B}{2}$$

Illustration No. 12

The height of triangle **D-C-E** is half the height of triangle **A-C-B** and the line **D-E** is half the size of rod **A-B**, therefore,

D'-E', the base of the "Repulsion Triangle" **D'-C'-E'** will be half as long. Vortex **C** will be "repelled" or "pushed" towards rod A-B with half the force, as described in Illustration No. 13.

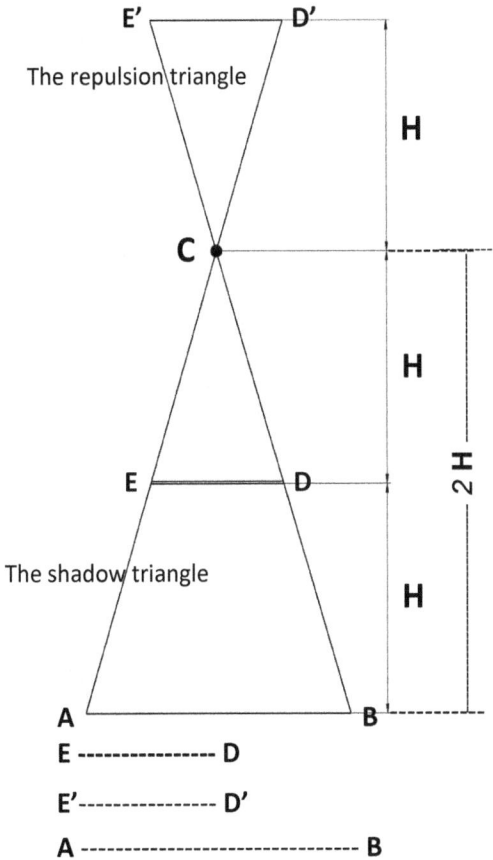

Illustration No. 13

In other words:

By increasing the distance of point C from A-B twice, the base of the Shadow Triangle is reduced to half. The base of the Repulsion Triangle is also reduced to half, so that the difference in forces is also reduced to half, and therefore, repulsion is also reduced to half.

If we distance body **C** from **A-B** three times (**3 H**), the repulsion force will be reduced by **3** times, according to the same principles.

If we move from a one-dimensional base, such as rod **A-B**, to a two-dimensional base, such as a square the thickness of a single atom, measuring 1m x 1m, or to a base whose sides are equal in length to rod **A-B**, we will have, in the language of Euclidean geometry, a "base" that equals A-B x A-B (Illustration No. 14).

If we place the material point **C**, at height **H** above the center of the base, we will get a pyramid with a square base, the area of which, S, equals the equation: S = A-B x A-B.

We will name this pyramid "The Shadow Pyramid," and will name the opposite pyramid, on the other side of the point **C** "The Repulsion Pyramid."

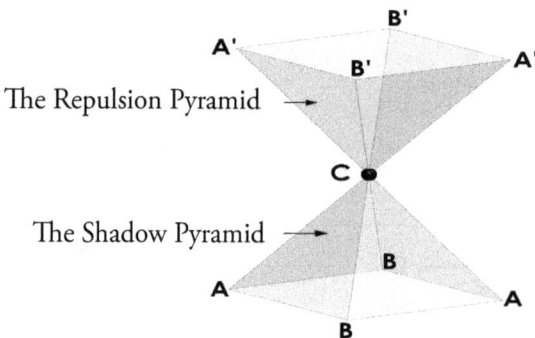

Illustration No. 14

If we raise point **C** to double the height (**2H**), we will have a new pyramid, with the same base, but twice as high.

We will mark a cross-sectional plane, parallel to the plane of the base of the pyramid at the height of **H** above the base (where previously point **C** was). We will also mark its points of intersection in the pyramid with the previous marking **D-E** on one side and **D-E** on the second side. Thus, we will have a square that is parallel to the base of the pyramid, the area of which, S_1, equals the equation: $S_1 = \textbf{D-E} \times \textbf{D-E}$.

Since the height **H** is two times smaller than the height **2 H**, the side **D-E** will be two times smaller than side **A-B** , and the area **D-E x D-E** will be four times smaller than the area of the base **A-B x A-B**.

If we raise point C to the height of **3 H** from its original height, the area of the base, which is at the distance of **H** under vortex **C**, will be reduced to a ninth (1/9) of its original size.

Similarly, if we raise point C to any height that is N times the height of H, the area of the base that is at a distance of H under point C will be reduced to $1/N^2$.

The considerations with which we have shown that the Shadow Triangle is equal to the Repulsion Triangle are also suitable to the Shadow Pyramid and the Repulsion Pyramid.

Thus, the sum of fracticles arriving at point C (in a unit of time) after crossing the base of the Shadow Pyramid is reduced by 4 when one distances point C twice as far from the base of the pyramid; or is reduced by 9 when one distances point C three times as far from the base of the pyramid, or, according to $1/N^2$, when point C is distanced N times away from the base of the pyramid.

In bodies with round, flat bases, the **Shadow Cone** will replace the Shadow Pyramid, and the **Repulsion Cone** opposite it will replace the Repulsion Pyramid, as described in Illustration No. 15.

In these cases too, the relations between the height and the area of the base will remain and the repulsion forces will decrease **in direct proportion to the squared distance.**

In the case of a "real" body which is thicker than a layer the thickness of a single atom, like in the previous examples, all we need to do is to think about it as a body with very many layers which have the thickness of a single atom:

> ➤ The repulsion force of every single layer decreases in direct proportion to the squared distance of that layer.

> ➤ The sum of the repulsion forces which are exerted on every body equals the sum of the repulsion forces of all its layers and it decreases in direct proportion to the squared distance, as demonstrated in previous examples.

> ➤ This conclusion matches what we know from experience and from quantitative formulations of the classic Gravitational Theory, derived by Newton from collecting data (Galileo, Copernicus and others).

> ➤ The Repulsion Theory enables us to understand why, for the flux of fracticles that comes from all directions and is partially blocked by bodies or masses in its way, the repulsion force **must change in direct proportion to the squared distance.**

The Repulsion Theory proposes a qualitative explanation alongside a quantitative explanation to the change of the force of gravity at an inverse proportion to the squared distance.[12]

---

12 Such as the change in the strength of the light that falls on the body from a regular source of light, which also changes at an inverse proportion to the squared distance of the body from the source of light (apart from a source of light such as a laser, for example). (The author)

The repulsion cone

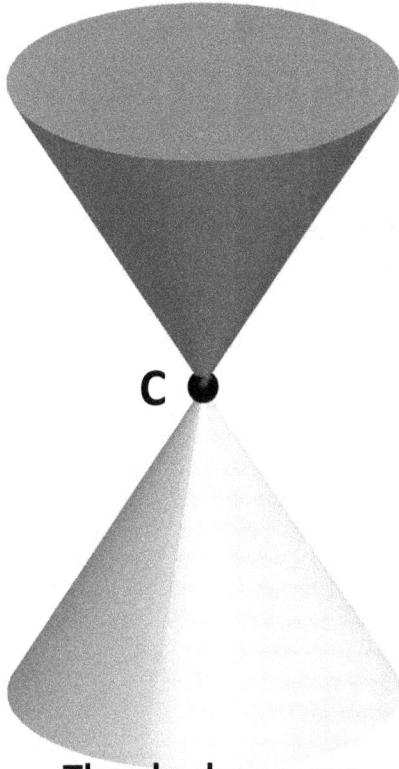

C

The shadow cone

Illustration No. 15

CHAPTER 3

- ➤ What is the Relation between Gravity and Electricity?
- ➤ Electric Wires and Electrons
- ➤ Cyclical Disturbances
- ➤ Magnetism

What is the Relation between Gravity and Electricity?

"The exception proves that the rule is wrong. That is the principle of science. If there is an exception to any rule, and if it can be proved by observation, that rule is wrong."

RICHARD FEYNMAN

Could there be some relation between "Gravity," according to the Repulsion Theory, and phenomena caused by the flow of an electric current through a conductive wire and the behavior of various bodies near it?

When an electric current flows through a conductive wire, electrically charged bodies will move towards it (or away from it) whilst other bodies will remain indifferent to the phenomenon, which seems as if the "direction of the operating force" is **perpendicular** to the direction of the current flow.

It is easy for us to accept the movement of bodies **"in the direction of the force"** that is exerted over them as understandable. However, it is difficult to accept the movement of a body **"perpendicular"** to the direction of the force exerted over it.

It is very difficult to imagine a body which has a certain force exerted over it, but which does not move in the direction of this force, but rather "prefers" to move in a different direction, perpendicular to the direction of this force.

In order to overcome this descriptive difficulty, several concepts were developed, including "force lines," "force field" or "field."

These concepts are useful in solving practical problems, but do not show any **comprehensible** causal link regarding the movement of bodies, from which we can deduce that there is a "force" in action, or that a "field" is present.

In order to present a perceptible explanation to the movement of bodies near a wire that conducts electricity, and how the repulsion theory explains these phenomena, we will use an analogy from *the Fable of the Mesh Tree Frogs*. The story goes as follows:

"Many years ago, there lived a strange tribe of tree frogs, *the tribe of mesh tree frogs.*

These tree frogs were made of delicate and strong mesh, with rather large holes. The jets of air which reached the mesh from all directions could pass through the holes without the tree frogs feeling anything.

The mesh tree frogs lived in a world that spread over a mesh tube, whose internal space was completely empty.

According to the fable, this tube started in a distant place on one side, and extended far into the horizon on the other side.

Small mesh trees are randomly grown on the surface of this mesh tube.

All around the mesh tube, air openings were situated at equal distances from each other and at a set distance from the surface of the tube.

Every mature mesh tree frog used to carry a large, mesh parasol which it spun very quickly over its head, and thus "protected" its head and back from the jets of air coming from above from the openings around the mesh tube.

The jets of air coming from below, from the other side of the tube, were hardly "disturbed" by the mesh wires from which the tube was made, and they created a pleasant coolness that flowed onto the belly of each mature mesh tree frog.

Scholarly mesh tree frogs thought of the idea of having rotating parasols, as they looked for ways to reduce the irritating gusts of winds on their smooth heads.

We could perhaps offer a loose translation of the frogs' consideration from 'Ancient Froggish' to our language, thus:

"When the parasol is at rest, the air molecules will pass between the wires with almost no disruptions. If the parasol spins around, its wide wires hit the air molecules which pass between them. These molecules will either get "swallowed up" by the wires or diverted from their course, but in any case, they will not hit our heads directly.

The more the rotational speed accelerates, the more molecules will be hit by the mesh wires, and the fewer molecules will touch our heads."

In order to examine the idea, the mesh tree frogs built an experimental device that creates a downpour of "a rain of peas."

Under the "rain of peas" they placed a parasol in the shape of a bicycle wheel, with metal strips connecting the hoop to the center of the wheel.

The metal strips were wide and thin, so that their narrow side faced the "rain of peas" and their wide side was perpendicular to the "rain of peas."

The fact that the stripes were narrow almost did not prevent the peas from hitting the floor beneath the parasol, where the number of hits was counted by a rapid counting device.

When the parasol spun, the wire stripes hit the peas and spurted them in all directions.

The faster the parasol would spin, the more peas would scatter to the sides and the fewer peas would hit the counting device that represented the heads and backs of the mesh tree frogs, which, in their opinion, "proved" the brilliant idea.

Within a short time, the idea became popular among the mesh tree frogs, who had, up until then, suffered from headaches which, as experts thought, resulted from the annoying jets of air.

Today, many years after that experiment, it is difficult to find a mesh tree frog without a revolving parasol...."

That is the story.

We shall now pause, in order to examine several aspects regarding the special world of the mesh tree frogs, before we continue with our story:

When, jets of air flow from the air openings towards the center of the tube, the mesh tree frogs will feel a pressure on their back from the air flow coming from openings above them, with an almost equal pressure on their stomachs from the air coming from the openings on the other side of the tube.

What would we observe if a "strong flow of peas" passed along the inner diameter of the tube, just, like water passes through metal or plastic pipes?

Would there be a difference between the pressures exerted on the surface of the tube?

Is there any connection or relation between the pressures exerted on the surface of the tube, and the speed of the peas inside the tube and their density?

What would happen to the mesh tree frogs living on the surface of this tube?

Each air molecule passing through the tube may hit one or more of the peas moving inside it, and the flow of peas will sweep away with it some of the air molecules, will divert the course of additional molecules and thus create a resistance to the passage of air molecules from one side of the tube to the other.

The stronger the flow of peas passing through the tube, and the more the number of peas passing through each point in the cross-section of the tube per time unit, the more chances there are for each air molecule to collide with a passing pea and get carried away with it.

The resistance of the tube (with the flow of peas) to the passage of air molecules through it will increase, and **the difference in pressures**, resulting from the flow of air coming directly from the openings above it and the flow of air entering through the tube and from the openings on the other side, will **grow**.

And the tale continues:

"One day, a noise was heard in the mesh tube, and a strong flow of peas started flowing through it…."

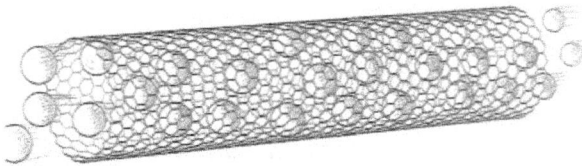

Illustration No. 16

As the noise of the peas in the tube was heard, the tree frogs tried to flee for their lives by jumping to the tree tops, away from the scary noise.

Even in their plight, the mesh tree frogs did not stop revolving the parasols above their heads, out of a naïve belief that this custom keeps them safe from any harm.

To their amazement, the mesh tree frogs discovered that the stronger the flow of the peas in the tube, the more difficult their hops became; a mesh tree frog that had previously reached the top of the tree with one hop, could now only jump a tiny bit, but would then return and fall down to the surface of the tube…."

Let us now stop and examine what happened to the terrified mesh tree frogs:

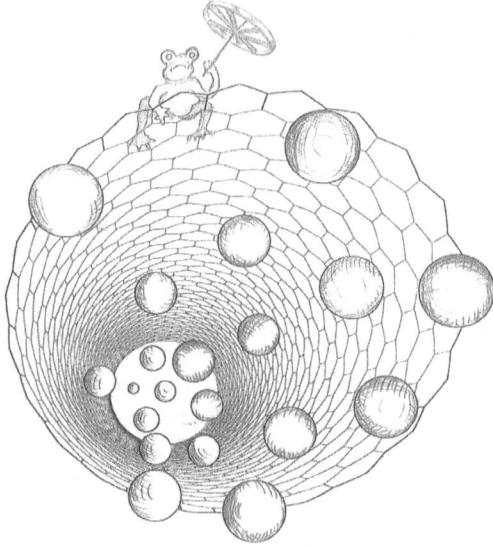

Illustration No. 17

The air pressure coming from the air openings above, which hit the spinning parasol is greater than the air pressure coming from below and passing through the tube with **the flow of peas passing through it.**

Each time a mesh tree frog tries to hop to the top of a tree, the difference in air pressures exerted on the spinning parasol, push it and the tree frog holding it back in the direction of the center of the tube.

Mesh tree frogs that hop high will feel a weaker pressure when they are high, as the "Shadow Cone" will have a smaller base, relative to its height, when they are at a great height.

These mesh tree frogs will finally fall onto the mesh tube, which will be most embarrassing for them.

Young mesh tree frogs that do not take parasols, or which do not yet know how to revolve a parasol, will barely be affected by the flow of peas passing through the tube.

These young mesh tree frogs will keep hopping to the tree tops, almost as if nothing had happened.

We, the observers, will easily distinguish the mature mesh tree frogs with the spinning parasols, which cannot jump high with the same ease as before, versus the young mesh tree frogs, without any spinning parasols, which keep hopping high as if there are no peas flowing through the tube.

With the flow of the peas, we may distinguish between two groups of mesh tree frogs:

> ›   Young mesh tree frogs that were not affected by the flow of peas.

> ›   Mature mesh tree frogs that were pushed towards the tube (or were "attracted to it," in more conventional accepted language).

What can we learn from the story of the mesh tree frogs, the peas and the air flowing from the openings around the tube?

If we treat the mesh tube, the air, the peas and the mesh tree frogs as one system, we may be able to notice the relation between the components of the system and how differences were created in the behavior of different kinds of mesh tree frogs:

> ›   If the tube were made of a transparent mesh, and the peas were transparent too, the behavior of the mesh tree frogs, as described, would seem "very strange" to us. This is since the reasons for that behavior would not be in our field of observation, just as the behavior of many bodies in our world is not clear or completely under-

stood, so long as essential, relevant information is not available to us.

➤ Can the falling of the mesh tree frogs towards the mesh tube, as a flow of peas passes through it, remind us of the behavior of metallic bodies when they are near a wire carrying an electric current?

➤ Could it be that the behavior of bodies that are close to an electric conductor is related to the same factor that also affects their gravitational "attraction"?

Can the special environment of the mesh tree frogs affect the conclusions we might draw?

ELECTRIC WIRES
AND ELECTRONS

"Our confused wish finds expression in the confused question as to the nature of force and electricity. But the answer which we want is not really an answer to this question. It is not by finding out more and fresh relations and connections that it can be answered; but by removing the contradictions existing between those already known, and thus perhaps by reducing their number."

HEINRICH HERTZ

Let us connect a conductive wire to a source of voltage and to an operating switch.

As long as there is no current in the conductor, the "effect" of the wire will not be felt by bodies near it.

Once we switch on the electric switch, and an electric current is flowing through the conductive wire, "electrically charged bodies" will start to be "attracted" to it. At the same time, other bodies, which are not charged, will not be affected by the electric current flowing through the wire and will not move towards it or from it.

If we compare the flow of electrons flowing in the wire or on its surface to the flow of peas that flowed through the mesh wire, and the "electrically charged bodies" to the parasols spinning over the heads of the mature mesh tree frogs, we may notice an analogy to the electromagnetic force. This force creates, as is widely understood, the "electromagnetic field" **around** the wire that is carrying electricity.

Electric charges or bodies carrying an electric charge will **approach** such a wire or will be "attracted" to it.

Bodies that will **not move** towards a conductive wire that carries an electric current, and which are not affected by an electromagnetic field, may be compared to the young mesh tree frogs without the parasols, which were almost unaffected by the flow of peas in the mesh tube.

Perhaps one may compare the electrically charged bodies to the mature mesh tree frogs' spinning parasols.

Electrically charged bodies (with "spinning parasols") near a conductive wire that has an electric current flowing through it will move towards the wire, perpendicular to the direction of the electric current (just as it happened to the mesh tree frogs in the tale).

Perhaps this effect happens due to the rapid spinning of the electrons around themselves, or any other fast movement that creates what we call "an electric charge."

What is this thing that we term "electric charge," and which we see as the reason for the behavior of the "charged" bodies?

> ‣ It is possible that an electric charge is not a "thing" carried by charged bodies, but a kind of spinning, rotation, or other movement of bodies in space.

> ‣ The motion of the bodies, whether a rotation around themselves or a vibration or another kind of motion that we cannot imagine, may create an "obstacle" to the passage of fracticles through them. This obstacle could be the spinning of electrons around themselves, which creates a disturbance to the free passage of fracticles in a defined area of space, such as the flow of air, which encounters difficulty in crossing the "area of spinning" of the parasols or the flow of peas in the tube that some-

what halted the passage of air molecules.[13]

> As the parasols that spin in the world of the mesh tree frogs are made of mesh, their resistance to the passage of air molecules will be **the smallest** when rotation is slow and **the largest** when rotation is fast.

> Perhaps the property that gives the electrons their "electric charge" is their self-rotation in space, vibration or any other motion that causes them to cross the flux of fracticles, which comes from all sides, and to create **a disruption** in it.

> The electric current, as is commonly assumed, is composed of a flow of electrons which move very rapidly on the surface of or inside a conductive wire. Perhaps we could compare an electric current with the flow of peas in the mesh tube. Its effect over bodies that are near it results from the disruption that it creates (due to its motion) to the passage of fracticles coming from all directions and **the violation of the symmetry** in the forces that they exert over the bodies that they collide with.

Perhaps, without an external flow of fracticles, the electric current in the conductive wire would not have caused the bodies to move perpendicular to the direction of that flow.

---

13   See chapter 3: What is the connection between Gravity and Electricity?

CYCLICAL DISTURBANCES

"It appears therefore that certain phenomena in electricity and magnetism lead to the same conclusion as those of optics, namely, that there is an Ethereal medium pervading all bodies, and modified only in degree by their presence; that the parts of this medium are capable of being set in motion by electric currents and magnets; that this motion is communicated from one part of the medium to another ..."

JAMES CLERK MAXWELL

Let us return again to the mesh tree frogs on the surface of the tube and to the flow of peas flowing in the tube, and examine a case where the flow of peas reaches in waves our observation point.

At a given moment, a strong flow of peas, passes through a certain point in the tube, and immediately afterwards, peas do not arrive at the same point. After a defined break, a new wave of peas arrives, after which there is silence, and so forth in a cyclical manner.

What will happen to the mesh tree frogs with their spinning parasols?

For a moment, the mesh tree frogs will not be able to jump high, but right after that, when the flow of peas in the tube stops, they will be able to hop again for a short while, until a new flow of peas will make it difficult for them to hop again, and so forth.

If we name the relatively small number of air molecules

moving from tube center onwards, perpendicular to the direction of the tube (after they have "crossed" the flow of peas) as a "disturbance" and name their return to a "regular" number as a "break," we may be able to see that a cyclical disturbance occurred in the flow of air passing through the mesh tube and perpendicular to it. This disturbance moves from the tube onwards, at the speed of the air flow.

The expansion in space of a cyclical disturbance to a flow of fracticles (which comes from the universe that surrounds us), as a result of a current flowing through an electric conductor, violates the balance of forces exerted over electric charges where the disturbance affects them.

This disturbance causes the charges to move from the place where they are towards the direction where the disturbance came from (fewer fracticles).

> ➤ The movement of a cyclical disturbance in the flow of fracticles may constitute what we term "electromagnetic waves."

> ➤ A seemingly essential condition for the existence of "electromagnetic waves" is: A flow of fracticles coming from the universe around us, as it passes through a body creating **a cyclical resistance** to this flow.

> ➤ A change in the direction of the flow creates a similar disturbance, and can be compared to a flow of peas coming from the side opposite to the original flow, while there is a break in the flow of the original flow.

> ➤ The time that passes between each disturbance defines the "length of the wave" of the disturbance.

> ➤ The difference between the number of fracticles coming from one direction to the one coming in the opposite direction defines "the size of the disturbance" or "the

height of the wave" or its "amplitude," and it depends directly on the "excess" of fracticles passing at a given period of time through a certain point in space.

➤ Perhaps without a flow of fracticles, there would not be any disturbances to this flow, there would not be any "electromagnetic waves" and there would not be any gravity (or "attraction force").

➤ What gravity, electric forces, electromagnetic waves, magnetic phenomena and other phenomena from the field of the atom have **in common**, can be described by "a universal flow of fracticles" coming from all directions in the universe and the disturbances caused to this flow by various bodies in its way.

➤ Two bodies **approach** each other when they create a disturbance to the flow of fracticles, and when the flow of fracticles between them is "thin" or smaller than the flow of fracticles beyond them (in the opposite direction).

➤ Two bodies **move away** from each other, when the flow of fracticles between them is "thick" or larger than the flow of fracticles coming from the opposite direction.

How is "a larger" or "thicker" flow of fracticles created between two bodies, which is different from the fracticles flow coming from the outside?

MAGNETISM

*"That the globe of the earth is magnetick, & a magnet; &
how in our hands the magnet stone has all the primary forces
of the earth, while the earth by the same powers remains
constant, in a fixed direction, in the universe."*

WILLIAM GILBERT

A magnet has a "strange behavior," the forces that it exerts
are not symmetrical in all directions and are not equal in all the
points on its surface.

The magnet "presents" an "indifferent relation" to materials
such as glass, plastic, wood and most metals, and a "special
relation" to iron and some of its alloys and to other magnets.

The "effect" of a magnet is powerful near two points on its
surface, and it weakens the more distant it gets from those points.
These points are usually termed "poles," and one is marked with
the letter "N" – North, whilst the other is marked with the letter
"S" – South.

Magnets have many shapes, but for clarity, we will examine
a magnet shaped like a straight rod, with its two poles at its tips,
marked as usual with the letter "**N**" for "northern pole" and with
the letter "**S**" for "southern pole."

When we bring a "northern" or a "southern" pole of a magnet
close to an iron rod that has no magnetic properties, the iron and
the magnet will draw close to each other and stick to each other.
Thus, one says that both poles of the magnet "attract" iron. This
phenomenon is widely known as "attraction," and we say that the

magnet "attracts" the iron, since iron does not "attract" other iron.

When you bring two magnets close to each other, one with a southern pole in front, and a second magnet with a northern pole in front, both magnets will "leap" and cling to each other at their poles, just like a magnetic rod clings to iron that is not magnetic. Thus, we say that both magnets are "attracted" to each other when they have different poles ("N" vs. "S") facing each other.

When two poles with an identical sign face each other (the northern pole of one magnet facing the northern pole of another magnet, or the southern pole of one magnet facing the southern pole of the other magnet), then the magnets "resist" when we try to bring them closer. As soon as we leave them, they will "run-away" somewhere far, where the mutual "resistance" will not be able to overcome the friction, as long as they are placed on a planar surface. Therefore, we say that poles with equal signs "repel" each other.

The Magnetic Field

If we spread iron sawdust on a planar paper, beneath which there is a magnet, the slivers of iron sawdust will arrange themselves in a way that is reminiscent of curved lines that have a higher density closer to the poles.

Most theories explain these phenomena by claiming that the magnet creates a "magnetic field" that causes these phenomena, or that it creates "curved force lines" along which it attracts or repels magnetic bodies.

What is that field and how does the magnet "create" it?

Do particles, waves or any kind of undiscovered particles move within this field?

What is in this field that can affect the movement of bodies such as iron or other magnets, but does not affect materials such as plastic, wood, glass and a number of metals?

How are "**curved** force lines" created?

Can these phenomena be explained if we claim that "a field" or "attraction" or "force lines" cannot rationally exist in a world of physics that is based on the principle of inertia?

If there is no "attraction," then perhaps there are no "force lines" or "curved force lines" along which the "attraction" or "repulsion" operate, and if so, how would we explain the source of these phenomena?

If the various bodies cannot attract each other, but can only be pushed towards each other by a flux of fracticles coming from around them, how can we explain that two magnetic poles with the same sign are "repelled" by each other or move away from each other, while two magnetic poles with different signs approach each other?

Perhaps the bodies are repelled by each other when the flux of fracticles between them is larger in number than the flux of fracticles coming towards them from other directions.

Perhaps a body with a special structure creates this phenomenon and "allows" more fracticles to pass through the body in one direction, but "does not allow" fracticles from other directions to pass through it "as easily" in the other direction.

Directional Preference

When the structure of a body enables fracticles to pass through it in a certain direction, the body will be pushed less in this direction, as most of the fracticles will pass through it without pushing it. When a body does not enable a significant part of the fracticles flux to pass through it, they will push it in the direction of their movement.

A body that "allows" a large number of fracticles to pass

through it in one direction but less so in the opposite direction will be subject to the action of **asymmetrical** forces in these directions, and to equal, or **symmetrical** forces in all other directions.

A body may be able to create a "directional preference" if it is composed of fracticles that are in a certain order, in linear or circular motion, or if it is near another body that halts or violates the symmetry of the flux of fracticles.

It is possible that the direction of a "directional preference" is derived from the self-rotation of some particles. If the "preferential" mechanism is related to the self-rotation of fracticles around themselves or around some other center, such as "vortexes,"[14] then changing the direction of the vortexes can cause a change in the direction of the "preference."

Directional preference probably depends on motion, and this depends on the motion generator, and so it will exist as long as the "preferential mechanism" operates, due to the location of a second body close to the first body. When we distance the second body, the preferential mechanism will weaken in direct proportion to the extent of the distance. The body will stop letting the fracticles pass in a directional manner and will go back to having no directional preference (just like most bodies).

An example of a mechanism of directional preference can be seen in the spinning parasols in "The Fable of the Middle East Tree Frogs" (Chapter 2) or the spin of electrons, as is currently assumed.

Perhaps the rapid spinning of fracticle vortexes "creates" the "difficult" points or "the disturbances" to the passage of other fracticles, a point from which the creation of "matter" may begin.

---

14   See Chapter 4: Mass or "Vortexes" (the author).

All this does not depend on the direction of the vortexes. Such a mechanism creates a directional preference, when most vortexes spin in the same direction, and when another body "hides" the flux coming from the location of the "hiding" body.

When vortexes spin in different directions and randomly, there is no directional preference, and the body will not behave like a magnet and will have no "magnetic properties" as we tend to term this phenomenon.

Reversal of Magnetic Poles

If the phenomena of magnetism are created due to a directional flux of fracticles, from one magnetic pole towards the other magnetic pole, we probably have a very slim chance of finding "monopole" magnets, since the flux of fracticles "enters" from one direction and "leaves" from another.

If we treat a magnetic rod as a body with a directional preference, we can imagine that it enables, for example, more fracticles to pass through it from the northern magnetic pole to the southern magnetic pole than in the opposite direction.

Regarding the direction that is perpendicular to the direction of the main flux (from one magnetic pole to the other), an **equal** number of fracticles will pass through the body in all directions.

Regarding different angles from the perpendicular to the "preferential axis," a larger number of fracticles will pass, the closer the angle of their entry or exit from the magnet is to the preferential axis, an axis in which there is the largest fracticles flux.

Force Lines

A sliver of iron sawdust placed on a planar paper sheet, above a magnetic rod, will be pushed ("pulled") with more force towards the magnetic poles compared to another sliver placed on the paper close to the center of the magnetic rod.

If an imaginary magnet would push the iron slivers away (rather than pull them towards it), we would expect the iron slivers to move away from it until the push would not be able to overcome the friction of the iron with the paper. Thus the iron slivers would stop at the point where the forces were balanced (repulsion versus friction). In such a case, we would see that the slivers tend to stay close to the center of the magnetic rod and further away from its magnetic poles.

However, with a regular magnet, the slivers are pushed towards the magnet ("attraction") in such a way that the forces pushing each sliver are stronger close to the magnetic poles, and weaker around the center of the magnetic rod.

Since the magnet is placed under a planar paper sheet, over which iron slivers are spread, the force pushing each sliver downwards (towards the paper) will increase the closer the sliver is to the magnetic pole. This will cause **an increased friction with the paper,** and each sliver will stop at the point where the forces are balanced.[15] Next to the magnetic poles, where the "directional flux" is great, the iron slivers will be crowded, and around the center of the magnetic rod, they will stop further away from it.

Perhaps the "curved force lines" that we see are imaginary lines that connect iron slivers with equal friction: lines of slivers of **"equal friction."**

It seems that "the magnetic forces" do not operate in the direction of these "curved lines," but **perpendicular** to them. Namely, with the direction of the greatest flux of fracticles ("most crowded") at that point.

---

15   The force of friction of the sliver with the paper sheet, equals, the force pushing each sliver towards the paper (downwards), times the coefficient of friction for paper (the author).

Perhaps the difference in the number of fracticles passing at different angles to the "preferential axis," together with the friction of the iron slivers with the paper on which they are placed, are "responsible" for the special arrangement of the iron sliver particles and for creating the **illusion** of curved force lines.

"The New Laws of Magnetism" [16]

At this point, we can formulate the "new" laws of magnetism slightly differently than usual, and, with these laws, search for new "openings" for an understanding of the magnetism phenomena:

1. Two magnets whose magnetic poles direct the flux of fracticles to each other will be "repelled" by each other, because "an excess of fracticles" in motion will be created between them.

  ➤ Two magnets whose magnetic poles direct the flux of fracticles "outwards" will have their magnetic poles reversed (N' and S' will switch sides), and they will pass the flux "inwards," so that in such a situation the magnets will be repelled by each other, just like in "Law No. 1."

2. When the magnetic pole of one magnet directs the flux to the other magnet, and the magnetic pole of the second magnet directs the flux onwards, the magnetic poles of the first magnet will be "reversed," and the two magnets will direct the flux "outwards." An area that is "thin" with fracticles will be created between them, and the magnets will be "attracted" to each other.

  ➤ The iron causes the magnet to direct the flux onwards, so that between it and the magnet, an area that is "thin with

---

16 The description of the magnetic phenomena is not complete and does not cover all known phenomena. Points such as "directional preference" and "pole reversal" need to be further developed, clarified, changed or improved (The author).

fracticles" is created, and the iron will be "attracted" to the magnet.

> Most matter does not cause a "directional preference" to be created in a magnet. Therefore, these materials will not be pushed to the magnet or from it.

> The fracticle theory predicts that under certain conditions, **the nature of which is still unknown**, any matter may become a magnet, if a mechanism of "directional preference" is somehow created in it.

> According to "The New Laws of Magnetism," magnetic poles are reversed under certain conditions (N' becomes S', and vice versa), and they direct the flux in the direction opposite to its previous direction. The reversal of the magnetic poles is required in order to describe the repulsion of equally signed magnetic poles that direct the flux of fracticles onwards, according to "Law No. 2."

> A reversal of magnetic poles occurs whenever two equally signed magnetic poles that direct the flux onwards, face each other and are near each other.

Can a similar phenomenon be found in a lab, or, in one of the experiments that nature "conducts" for us?

The Earth as a Magnet

Planet Earth, where we live, reverses its magnetic poles irregularly once every hundreds of thousands of years or once every millions of year, as one can learn from the examination of basalt rocks that have magnetic properties.

During a volcanic eruption, basalt contains "ferromagnetic" matter, in the form of molten lava.

The lava enables tiny magnetic crystals in it to arrange themselves in the direction of Earth's magnetic poles, so that the magnetic pole **N'** in the tiny magnet will face Earth's southern magnetic pole, and the magnetic pole **S'** will face Earth's northern magnetic pole.

When the lava cools down, the tiny magnets "freeze" and remain the way they were arranged when the lava was still soft (molten).

By using different techniques, one can determine the time when this process took place. Furthermore, we can see that the magnetic poles in the basalt face northwards at a certain period and face southwards at a different period, according to the condition of the magnetic poles of Earth at that time.

This "story" of the basalt could possibly indicate that in these periods, a large astronomical body with very powerful magnetic properties passed **close to Earth** and caused the reversal of magnetic poles of planet Earth...

Geologists can determine these periods quite accurately. With them, it may be possible to realize how lucky we are that such a body or bodies may have passed "nearby," but did not hit us directly.

Will we be so lucky in the future?

It seems that a relatively small numbers of "laws" enable us to build a rather good (though still incomplete) description of the "strange behavior" of magnets, and to connect this phenomenon, through fracticles – to gravity, magnets, electrical phenomena and to electromagnetic waves and forces.

CHAPTER 4

- ➤ Force
- ➤ The 'Field,' A Reality or an Intellectual Model?
- ➤ Mass, or Vortexes?

FORCE

"Many powers act manifestly at a distance; their physical nature is incomprehensible to us; still we may learn much that is real and positive about them, and amongst other things something of the condition of the space between the body acting and that acted upon, or between the two mutually acting bodies. Such powers are presented to us by the phenomena of gravity, light, electricity, magnetism, &c. These when examined will be found to present remarkable differences in relation to their respective lines of forces; and at the same time that they establish the existence of real physical lines in some cases ..."

MICHAEL FARADAY

"It is evident, therefore, that our- physical views are very doubtful ; and I think good would result from an endeavor to shake ourselves loose from such preconceptions as are contained in them, that we may contemplate for a time the force as much as possible in its purity..."

MICHAEL FARADAY

The Forces in Action

When a body, for some reason, diverts the flux of fracticles coming towards it from all directions, so that more fracticles will move towards another body near it, this could cause an "excess of fracticles" or repulsion between the two bodies.

When, for some reason, the rate of emission of fracticles is greater than the rate of their absorption, an "excess" of fracticles will be created between the bodies and will cause the two bodies to be repelled by each other.

If the flux of fracticles between two bodies is denser than the flux of fracticles coming towards them from all other directions, the bodies will be repelled by each other and will distance themselves from each other.

All bodies are "able" to halt the flux of fracticles to some extent, but it seems that not all bodies are "able" to divert the flux of fracticles coming towards them or to create an excess of fracticles.

Therefore, we may not observe attraction or repulsion in all bodies, as we see them between magnets or between a magnet and ferrous metal, namely a reaction of "charged bodies" or the creation of "electromagnetic" waves.

Perhaps the possibility of diverting the flux of fracticles results from the fast rotation of small particles (such as electrons) or from their special organization within matter, as mentioned previously about magnetism.

The fast rotation of the various "basic particles" may cause a disturbance to the passing of the flux of fracticles, such as the disturbance created by the "revolving parasols" in the world of the "mesh tree frogs" (in Chapter 3).

The disturbance created can stop or "swallow up" the flux of fracticles or divert it to a different direction than the direction in which it was moving before it collided with this body.

Perhaps the fast rotation of the basic particles from which the atom is composed of according to the "standard model" assumptions, such as electrons (leptons) or protons, neutrons or smaller basic particles (quarks) that compose it, enables, the swallowing up or diversion of the flux.

As the rotation (or any other movement) of particles grows

faster, the "forces" that bring together or distance the bodies from each other become greater.

Elementary particles that are motionless (or almost motionless) barely cause any disturbance to the flux of fracticles, and therefore, they only create small "forces" between the bodies.

A fast rotation of the elementary particles, or any other motion that is difficult for us to imagine, creates a greater disturbance to the passing of fracticles.

Today it is common to attribute to that disturbance the labels of: electric forces, magnetic forces or "other forces" within the atomic nucleus, which are believed to be much stronger than "gravity."

What, then, is a force? Can we attribute an independent physical entity to a state labeled as being a "force"?

Force

When a body causes another body to move towards it or away from it, we usually say that it exerts "an attraction force" or "a repulsion force" over it.

What is that "thing" that is "exerted" by one body over another, which we name "force"?

Is "something" indeed being exerted beyond the bodies' properties to halt or to divert the flux of fracticles coming towards them from all directions?

Is the concept of "force" a real thing, which can cause a body to move?

Can force exist without motion?

Or more specifically - can one body affect another body without "something" moving near them or between them?

Can remote impact or remote influence of one body be

exerted over another body, which would cause one of them to move, without any previous motion of a particle, wave, body or any "thing" that could move between those bodies?

When we see a motion of bodies that is different from constant speed in a straight line, we attribute the motion factor to some "imaginary" value which we term "force," as we are used to utilizing the letter 'x' to represent a numerical value for an "unknown variable" in algebra.

'**X**' itself has no physical meaning, even if it describes 100 tons of lead. The lead, unlike 'x', represents some physical reality, whilst the 'x' represents just a way of thinking or an intellectual model.

Perhaps the meaning we should attribute to lead is the possibility of "creating" a difference in the number of fracticles coming towards it and passing through it from one direction, compared with the number of fracticles coming towards it from the opposite direction.

The difference in the number of fracticles causes lead to be heavy and to move towards the center of the earth. This is just as it also causes the motion of all other bodies in the universe, from the very smallest known to us up to the largest and most distant galaxies observed with the aid of the equipment we have developed.

Perhaps the concept of "force" has no physical meaning. Perhaps force is not a real thing leaving body **A** and causing body **B** to move in "some mystical way" in the direction of that "force" or in the opposite direction, namely, to move away from or move closer to body **A**.

The concept of force assumes the existence of a medium that exists between the bodies and causes them to transfer motion from one body to another in an inexplicable way. Otherwise, it assumes the existence of "force carriers" that move in contrast to the "accepted" laws of physics and **that particularly violate the principle of inertia.**

One can continue using the concept of force in the same way in which we use '**x**' as an "unknown variable" for calculations, if we just distinguish that we are discussing a figment of our imagination or an intellectual model, rather than an actual occurrence.

There are perceptions in physics that perceive force as part of a "field" that sprawls from near a body until the edges of the universe, whilst the force of this field weakens somehow as the distance increases.

What does the concept of "field" encompass?

Does a "field" have a physical meaning?

Should we see a "field" as part of some reality, or is it better to refer to it as an intellectual model that enables us to solve practical problems by using an abstract value, just like '**x**' in the algebra equation?

What is a field composed of, and to which category should we ascribe it?

The 'Field', a Reality or an Intellectual Model?

"Consider then a mass of matter or a particle for which present purpose the sun will serve and consider a globe like one of the planets as our earth either created or taken from distant space and placed near the sun as our earth is the attraction of gravity is then exerted and we say that the sun attracts the earth and also that the earth attracts the sun. But if the sun attracts the earth that force of attraction must either arise because of the presence of the earth near the sun or it must have preexisted in the sun when the earth was not there ..."

Michael Faraday

When the motion of one body is different than a constant speed in a straight line, due to the effect of another body near it, we tend to say:

"The body is in a field created by the other body."

When bodies which are close to earth fall towards it, we say:

"They are in the earth's gravitational field."

To the same extent, earth is in the "gravitational field" of the sun, which, in turn, is in the gravitational field of another celestial element (galaxy) etc.

When we think that a body moves under the influence of magnetic or electric forces, we say, "The body is in a magnetic field, or in an electric field."

What is that "field" and what occurs in it, which can affect the motion of bodies, which allegedly are in it?

Can there be a field with remote influence, when nothing moves in it?

Can a field that has nothing moving about within it divert bodies from their linear motion at a constant speed?

Does a field have any kind of physical meaning, or is it, just like "force" and 'x', merely an intellectual model, rather than a real thing?

Does the concept of "field" have a "supernatural" influence over bodies in it, as a result of which they will change their location relative to other bodies, even without any "direct collision" by any kind of fracticles?

Perhaps the best answer we can offer in response to all these questions should be **a negative answer!**

The concept of 'field' probably expresses our lack of knowing or lack of understanding of **the source of motion** of bodies, resulting from their proximity to other bodies.

Perhaps "field" as a physical concept cannot exist, without some kind of fracticles moving about within it, which would cause bodies in this flux ("the field"), to move as they do according to our observations.

Could there be a motion of material bodies that begins by itself?

It is very difficult to accept as reasonable the view claiming that the motion of a body can start by itself, without there being any other initial motion.

The motion of one body can cause the motion of another body just by colliding with it. That is, by pushing or repelling it, or in the language of physics, by transferring momentum.

One might also be able to expand this perception to the

field of quantum mechanics, on condition that it does not see the source of movements in the bodies themselves, but in an **external** factor, such as fracticles.

A field, without a "flux of fracticles," or anything else moving about inside it, cannot be considered a physical concept, as it describes a static state rather than some **occurrence**.

A field presents a perception of a "reactive space" without the cause of the reaction as a possible explanation for the movement of bodies within it.

We measure movements and explain the results as if their cause lies in the existence of some medium called a "field."

Perhaps the field is no more than an intellectual model created by man, and as such, it can continue to serve us in the practical solution of problems in the world of physics. However, we should avoid "attributing" to it physical properties or seeing it as "a reality," or part of a measurable external occurrence.

If a field is no more than an intellectual model, what are the masses whose movement we follow?

Mass, or Vortexes?

"General Relativity — the physics of curved space-time, including the laws by which mass-energy and momentum curve space-time, and by which that curvature influences the motion of matter and influences the classical laws of physics..."

Roger D. Blandford and Kip S. Thorne

What is a **mass**, to which we attribute properties of solidity, strength, density and so on?

Does such an entity "really" exist?

Should we stay with a view that sees the mass as "the essence of matter," or should we prefer to treat it as "an area" or "a domain in space" that violates the symmetry influx of fracticles coming to this area from all directions?

What might happen to a single fracticle moving at the speed of light, in a straight line in space, which is encountered in an area that it "cannot" cross in a straight line that would be the direct continuation of its movement from up until then?

Let us visualize this fracticle as a fast swimmer that gets caught up during his swim in a fierce whirlpool, which makes him spin around in circles and thus "cuts off" his original swimming direction. Such a swimmer may slow down his swimming or stop, or change the direction of his swimming, and if he is not a good enough swimmer, heaven forbid, he will probably sink into the depths, get carried away or drown in the whirlpool.

Could a somewhat similar thing happen to fracticles, which

move about in the universe, and come across a "domain in space" that "disrupts" their movement, which we term as mass?

Could it be that the thing that "cuts off" the movement of the fracticle in space is a vortex of fracticles that prevents the fracticles coming towards it to pass freely, and thus, cuts off their undisturbed movement through the empty space?

We previously assumed that the flux of fracticles coming from the universe around us is equal from all directions. Thus, a flux coming from one direction passes through a flux coming from another direction without any mutual disruption or effect, such as light rays coming from different directions and passing through each other without any measurable effect. Such a possibility would seem reasonable if we assume that the distances between the fracticles are much larger relative to the "size" of the fracticles moving in straight lines through the empty space.

If there is an "overcrowding" of fracticles in a certain region in space, so that the flux of fracticles coming to that area cannot pass through it freely, a single fracticle might collide with other fracticles repeatedly, and thus change its direction of movement each time. These collisions would also affect the motion of the fracticles colliding with that fracticle. We could view the region where these collisions occurred repeatedly as "a disturbance region" with regard to the free passage of fracticles.

The entire "disturbance region" is also slightly diverted into the direction of the original route of that fracticle and at a rate relative to the number of fracticles in that region that were directly or indirectly "hit" by it.

"An overly crowded region" could be a local vortex of fracticles, the outer part of which tends to gather in the center, or move from the center to the perimeter.

We tend to imagine a vortex as a whirling current of water,

close to the surface of the water, in a two-dimensional way, so that the center of the vortex moves downwards in another dimension. Perhaps, when we try to describe miniscule fracticles moving about in a vortex, the motion can be **from the perimeter** of the vortex **to its center** or **from the center** of the vortex **to its perimeter** in a three-dimensional, spatial way.

A three dimensional, "spherical vortex" can be a "building block" for the creation of a **mass**.

Perhaps what we term "mass" or "matter" can be described as "local density" of a relatively large number of fracticle vortexes, which compose the elementary particles of the atom. Perhaps mass or matter, are no more than a large "collection" of fracticle vortexes, which create the elementary particles, such as electrons, protons, neutrons and others. These comprise the atoms that build molecules, from which all matter known to us is formed.

Describing vortexes as "disturbances" to the flux of fracticles could expand our understanding of the concept of matter. This assumption could possibly lead us to recognize that in the entire universe, there are only fracticles, which move about in straight lines, and vortexes or "local disturbances," which are also composed of "clusters of fracticles." These move about in a way that creates a "disturbance" to the passage of fracticles that move in straight lines.

Miniscule, three-dimensional vortexes, composed of fracticles, could pose the only block in the universe to the passage of fracticles moving about in straight lines.

Perhaps the fracticles, as we assume them, are also composed of small vortexes, made of even smaller vortexes.

There is currently no experimental method available to examine this assumption, and perhaps it does not matter whether vortexes are as described, or have a similar shape or a completely different shape.

A "curved" or "circular" motion in the world of fracticles is composed of fracticles moving in short, straight lines, which change direction each time a fracticle hits them from a different direction.

When observing this from a higher scale, one can see movement in a number of short, straight sections changing their direction, **as if** forming a circular motion, which enables the creation of vortexes or any other motion that appears to us as if it is not following a straight line.

The distance from which we examine the motion, creates the illusion of a circular motion, although in the world of fracticles there may only be linear motions accompanied by a change of direction, when two or more fracticles collide.

We have no idea what the "real" nature of vortexes is, perhaps, apart from the assumption that they somewhat halt the flux of fracticles coming towards them.

When the number of fracticles hitting a vortex is identical from all directions, the vortex will continue whirling at the same speed. If the vortex is not at the center of a "finite universe," then the number of fracticles hitting it from different directions will be different, depending on its location relatively to all the other "masses" around it.

Vortexes that are at the center of the universe will be "hit" by an equal number of fracticles from all directions, and their rotation speed will remain constant. All other vortexes will be "hit" by various numbers of fracticles coming from opposite directions, and this asymmetry will slow down or speed up the whirling of the vortexes, as the result of uneven hits, which are repeated many, many times. The speed of fracticles close to the edges of an accelerating vortex may eventually reach a critical value, which would cause them to become detached from the vortex and would "transfer" them to a linear motion in some direction in space. These fracticles would move about until they would, at some point, hit another vortex and cause it to increase its speed and "release" additional fracticles, which would also move in straight lines.

The process of increasing speed and "releasing" fracticles can cause the size of the vortexes to remain more or less constant, in terms of the number of fracticles and the speed of rotation.

The speed of the releasing of the fracticles, or the critical speed by which fracticles are "emitted" from the system, is perceived by our senses or our equipment, as the maximal speed in our world or as **"the speed of light."**

The value of this speed may derive from the size of the vortexes and a critical speed of rotation, beyond which the collision of fracticles coming from outside the vortex "cannot" hold the fracticles within it.

Perhaps fracticles that have already reached their critical speed "leave" the vortex and start their "journey" in a straight line at the speed of light, towards empty space and other vortexes.

Without the movement of fracticles, we would not observe the inequality caused by the disturbance to motion, and this disturbance would also not be able to exist without any motion.

So, perhaps the essence of a mass is **motion**.

If the essence of a mass is motion, there may not be a physical reality to the term "mass at rest," since mass has not even a very small part that is at rest.

The term "at rest" must derive from our daily lives, but is not suitable to the "stormy" world of fracticles.

It may be better to compare our world to armies of flies, as mosquitoes fly around every fly, each one of them surrounded by smaller entities ... and so on.

In this "stormy world," a few single bees move about in straight lines, until they unintentionally stumble upon one of the many "armies" spread in space.

There is no rest in our universe.

Perhaps light, which travels between the stars, is also composed of those fracticles, and therefore also has no mass at rest.

Motion probably "creates" the "mass," giving bodies their volume, solidity, softness, appearance and color or any other property that we perceive through our senses.

With no motion, there would be no universe and no creatures that would notice it.

Perhaps all natural phenomena can be simply described through the use of fracticles and motion, without any need for additional properties, apart from "blocks" or "difficulties" in the way of the flux of fracticles.

Perhaps everything we see, or experience in any other way, is derived from the flux of fracticles coming towards us from all directions, and within this flux, from those clusters of fracticles "disturbing" the "free flow of the flux."

Maybe this flux and the disturbances in its way create all the things we witness in our world.

Perhaps their arrangement, or the special structure of these "disturbances" and the way in which we interpret the effect they have on our senses, create the complexity and beauty of the universe, out of a remarkably simple collection of miniscule fracticles moving about in empty space, apparently at the speed of light.

Furthermore, we too, if I may add, are no more than "a collection of fracticles and disturbances" to the passing of this flux.

CHAPTER 5

- ▸ Speed
- ▸ Time
- ▸ Limitations to Human Thought

SPEED

*"No eye, wherever in matter it might be placed, has a sure criterion for telling from the phenomena where there is motion, how much motion there is and of what sort it is, **or even whether God moves everything around it, or whether he moves that very eye itself.**"*

WILHELM LEIBNIZ

If the entire universe is composed of fracticles and motion, how shall we define **motion**?

It is customary to define motion quantitatively through speed or velocity.

Speed is defined as the distance a body passes in a unit of time, or distance divided by time.

Let us try to examine "distance" and "time," two components of speed, to see whether speed represents an actual physical reality, or whether, like most concepts, it expresses an "intellectual perception" or "interpretation of reality," but not a measurable, physical phenomenon.

For example, the average distance between Earth and the Moon is approximately 284,000 km, and as far as we know, there is hardly any matter, substantial matter, such as air, gas or any known particles between the two.

There is probably no matter between Earth and the Moon that could be measured, counted or weighed in any way known to us. Thus, the concept of the "distance" that exists between Earth and the Moon does not represent anything "real."

The kilometers that we count are nothing but figments of our thought or imagination. Therefore, we would need a complex sentence, or several sentences of explanation, to define or describe them. One cannot "show" kilometers in any way, without using an image or an example.

One might consider distance as a "comparative value," which relies intellectually on "real," defined bodies, such as Earth, the Moon or the "standard meter," presented in Sèvres, near Paris.

The use of a "comparative" concept includes within it the conductor of the comparison, who compares in a "logical," intellectual way between existing bodies and sizes.

If, for instance, the Moon or Earth were to disappear, would **"the distance"** between them still remain?

One can "see" that without the bodies (Earth and the Moon), distance has no meaning, and it will disappear if they both disappear, or even if just one of them disappears – since distance itself has no independent physical existence.

If Earth and the Moon were to remain in place, but without the human population, would the distance between them, still remain?

Without people, there will be no comparisons, as there would be no one found to conduct them. Thus, since distance expresses "a comparative value," **distance will have no meaning.**

The "disappearance" of the human race as an "intellectual exercise" may also serve us as a way to distinguish whether a certain thing exists on its own, or whether it is a result of observation, thought and transformation of the results of the thought process into a "concept" that exists in the language and conversation held between people.

Most of our communication is conducted by using concepts

or images, rather than through the objects themselves.

Therefore, this form of communication makes us attribute properties of "real objects" to those concepts or images.

"Distance" can be seen as a result of a comparison between two concepts or more. In the example we chose, the comparison shows how many times one can place a "standard meter" in a kilometer and how many kilometers one can place, one after the other, **through thought only**, between Earth and the Moon.

This entire comparison takes place in our minds or our imagination or in our measuring equipment. It **does not exist** between Earth and the Moon.

Comparing two "length" values may be done in thought by placing both values beside each other and stretching an imaginary line from "the top" of the first value towards the second value, parallel to the plane in which they are placed.

If this imaginary line passes over "the top" of the second value, we say that this value is smaller than the first. If it hits the top of the second value, both values are equal, and if it passes under "the top," the second value will be larger than the first.

One may see the comparison as a form of thought that is not a direct result of reality achieved through our senses, but a result of "using the software" with which we interpret the information we receive.

The comparison has no independent, physical existence and it, like many other concepts, will not exist without any thinking, comparing creatures.

If the comparison derives from language, which is a system of human agreements, then "distance" would also be "a product of agreement," rather than **an actual thing**, which exists without thinking and talking humans.

If distance does not exist as an independent, physical entity, what about the concept of **time**, the second component of speed?

Does time have an independent existence, which does not depend on the existence of the human species?

Does time represent an independent physical entity, or perhaps...

TIME

"Not only do we measure change by time, but time by change, because they are defined by one another."

ARISTOTLE

"Time is nature's way of keeping everything from happening at once."

JOHN WHEELER

We say: "We exist and time passes."

The Chinese say: "Time exists and we pass."

For many, the use of the term "time" is a vivid, daily matter, so that we tend to attribute it with physical properties when it appears in our thoughts or in our imagination, and we see it as a "thing" that passes, moves or changes.

The expressions "I have the time" or "I do not have the time" express the possibility of accumulating or lacking something real.

"Time passes quickly" or "time passes slowly" are usually attributed to the pace of events in which we are, as time serves as a real point of comparison in a changing world.

The expression "time is money" provides a sense of "actuality" to the idea that time is a thing that one can receive money for, and thus we should refer to it with the respect it deserves.

Does time really exist?[17]

> Does time have an independent physical entity?

> Can one measure, count or weigh time?

> Can time exist without motion?

> Can one sense time in one of our senses or measure it with one of the devices created by man?

> Does time have a direction?

Finding answers to these questions could clarify the meaning of this "thing" called "**time**."

If we look at a clock with hands and dials and follow the movement of the big hand for five minutes continuously, will we observe that five minutes have passed?

Probably not!

Our eyes see the movement of the big hand moving from one digit to another, and we notice that this hand has passed a certain distance between two adjacent digits on the dial of our clock.

Is the distance that the big hand passed in our clock identical to the distance that the big hand passes in another clock?

As we know from our experience, the tips of clock hands in different-sized clocks pass different distances in **equal periods of time**.

The big hand in a clock with a large dial will pass a greater distance than the big hand in a clock with a smaller dial. Thus, the distance passed by the tips of clock hands in different clocks cannot express the time that passed since they left one digit and reached its neighboring digit.

---

17   For further reading on 'time', we recommend Henri Bergson's book: *Time and Free Will: An Essay on the Immediate Data of Consciousness* (*Essai sur les données immédiates de la conscience*, 1889) (editors).

134 | *Attraction – Secrets of Gravity*

If the clock hands in a clock with a large dial pass a greater distance than the clock hands with a smaller dial at the same time, the tangent speed of the edges of the clock hands will be different in different sized clocks, and so this cannot be a way to measure the time that has passed.

In contrast, the angular speed in most clocks is identical, namely, the number of degrees that the large hand passes between one digit and its neighbor will be identical in most clocks. Two clock hands in two different clocks will together go from one digit to the next and pass an equal number of degrees, although the tangent speed of the clock hands is different.

With our eyes, we can see the movement of a clock hand from one digit to the next, and we can observe that clock hands on different clocks pass **different distances** at the same time.

We see that clock hands on different clocks pass **an identical number of degrees** at the same time.

If clocks were built in the shape of a straight ruler, which starts with the number "0" and ends with the number "24," we would clearly notice that the clock hand moves faster on the long clock than on the short clock.

Could it be that we only observe motions, and that it is only the "interpretation" that we give to these motions that means time for us?

We cannot **see** time, although the tracking of passing time is done with our eyes.

We cannot **hear** time, although we can hear the call of the cuckoo clock, the faraway ringing of the church bells, or the annoying beeping of the alarm clock calling us to rise to work, instead of sleeping peacefully.

The call of the cuckoo clock, the ringing of the church bells, and the beeping of the alarm clock are agreed signs, which symbolize to us time or transition or the duration of time. These signals, without any specific interpretation, are at most "noise" to us, rather than time.

Just as the eye perceives the motion of the clock hands and the brain "translates" this motion to a concept of time, so does the ear perceive the ringing of the bells or the call of the cuckoo clock and the brain "translates" these signals as time.

We cannot see or hear time, though through eyesight and hearing, our brain "creates" the concept of "time."

In digital watches, we do not observe the passage of time, but rather we see the digits change. These digits are symbols, and our interpretation attributes to them the concept of time.

Perhaps one might say that "time" as a concept is an agreement or a symbol.

Perhaps time is not located in physics or in the world that surrounds us, but in thought, in language or in our "agreement" as part of the culture in which we live, and if we look around, we may find that not all cultures have a concept of time.

Perhaps the reason we cannot see, hear or feel time is simple: **time has no existence as an independent physical entity**.

Time exists as a perception, as a concept, as "conversation" and not as part of any reality. Perhaps time, a human creation, **does not pass because it does not exist!**

If we manage to seal a room completely, so that something real like matter, a particle or a wave will be unable to enter or exit this room or pass through its walls, and we leave a working clock in the room, we will see that "time will have passed" according to the clock, when we next open up the room.

Sealing up the room does not interfere with the "occurrence" of time within it, since time has no objective existence. It is not created outside the room and then permeates in through one of the processes known to us or that we are due to discover.

Perhaps time is just an "intellectual model" represented by motion.

We are able to measure time passing in the sealed room through the motion of the clock hand or hands, or the changing of the digits on a digital clock, which are a result of the internal motions in the mechanism of the clock.

Perhaps **memory** "creates" time for us, through comparison.

We remember the position of the clock hands at the beginning of the experiment and we see their position at the end of it. Thus, by comparing the "before" and "after" photos through our brain, which processes the data, we "create" the concept of time.

Therefore, time is an outcome of memory, and memory is a human property - so time is probably a human creation rather than "a natural phenomenon."

Let us imagine a room that is completely sealed, in which there is a clock, which sounds a short beep every minute. This clock has no hands or dial, and only a man who hears the beeps will notice its activity and will know, by counting the beeps, how much time had passed since the start of the experiment.

However, if the room where the clock is located has thick, sealed walls, we will not be able to hear any sound or beeps coming from it.

What will we know when we open up the room - did time pass there?

When there is no basis for comparison, we cannot know whether time has passed in this sealed room.

If we manage to create a sealed room where there is no movement, namely, there are no molecules, atoms or free electrons, and no light or any other wave from the field of electromagnetic waves or other waves enters it, in such a room, **time will not pass.**

If we place (an imaginary) clock in this room, which has no motion of molecules, atoms or electrons or anything else that can move, and there will be nothing that moves in the room, there will be no change in the condition of the clock, neither inside or outside, even after millions of years have passed outside that sealed room. In such a room, no change will take place, and no time will pass!

One can also see that without **motion and without memory, there is no time.** Without motion, we cannot observe the change in the external world, and without memory, we will not be able to observe motion.

Perhaps motion creates all the observable phenomena, and memory enables us to notice them.

Perhaps it is motion that causes the existence of objects, as well as the occurrence of all phenomena in the universe.

Time is the creation of our mind, or one of the possible ways to perceive the concept of motion, and not a real, physical thing, which invades everywhere and accompanies every phenomenon, as many of us may think.

If distance does not exist as an independent physical entity, "space" cannot exist as such an entity either. If both time and space have no such entity, then the concept of "space-time" has no independent physical entity...

From this perspective, the space-time model will also be merely an intellectual model, rather than a description of a real occurrence in the world of matter and motion. One can attribute to space-time any desired property, on condition that it fits the observable facts.

"Curved space-time" is one of the possible ways to describe the mutual influence of adjacent masses, but possibly not the only, or best possibility.

Time has **no direction**, but we, as humans, can attribute to it a direction, just as we toy with the ideas of forward or backward "time travel."

If distance does not exist as an independent physical entity, and time does not exist as an independent physical entity, then speed, which is the division of distance in time, will not exist as an independent physical entity.

If speed as a measure for motion, does not exist as an independent physical entity, what still does **"exist"** in reality?

Limitations to Human Thought

"It is well known that all sciences are based ultimately on pure rational reason. However, even though science is based ultimately on rationality, the 'empirical' sciences, such as physics, adopts, some universal principles based on experiential evidence."

Immanuel Kant

Can one know for sure what exists in reality and what exists as an intellectual model, or what exists in nature independently of man and what exists in language that would not exist without man?

What are these limits of human thinking?

Can the "thinking mind" discover the limits to its thought?

Can the discussion on the limits to our thought contribute to our understanding of natural phenomena?

One may say that a significant part of phenomena that we, as humans, attribute to nature, are not "natural phenomena," but "intellectual models," such as time, distance, speed, mass, energy, force" and so on.

These concepts "live" in language, but have **no independent existence without being accepted in language.**

As long as we are aware of the fact that most of the concepts that we use derive from language, their contribution to the advancement of thinking and experimentation is great. However, if we attribute these concepts with objective meaning or an **independent** physical existence, the advancement of thinking and experimentation may stop.

In order to distinguish between "things as they are" in nature, and our concepts, which exist in human language, we should examine how we humans distinguish between them.

Can one examine how we perceive external information and how we interpret it, by using images from computer language?

If so, then what is the nature of the "input" that we receive from the outside world and what is the nature of the "data processing" conducted in our brain?

It is commonly thought that we receive external information through the senses: Sight, hearing, touch, smell, taste and perhaps also through other senses, whose existence we may not be aware of.

We receive most of the information through our sense of **sight**. Thus, we should examine how we receive information through sight, and then consider how we receive information from our other senses.

Our eyes are the organ that perceives the information defined as "eye sight" as light coming from the sun or from other sources of light, hits various objects, passes through them or is reflected from them to our eyes.

The human eye contains a sophisticated mechanism, which allows the light reaching it to pass through the cornea, aqueous humor and the pupil, and to focus, with the aid of the lens, on the retina, in the back section of the eye ball.

Schematically, there are millions of photoelectric cells in the retina, onto which focused light rays fall, which are connected to tiny nerves. These tiny nerves come together and create the

nerves of the eye, which transfer the stimuli created by the light (in both eyes) as **electric pulses** to the brain. This is accomplished through a complex, direct and crisscrossed electrochemical route, which leads to the occipital lobe of the cerebral cortex, where the **"vision center"** is probably located.[18]

The brain, as we assume, "processes" the signals coming to it from the nerves and "presents to us" the results of the processing of the signal as an image.

Describing vision in this "schematic" way may already demonstrate, at this stage, that our brain **cannot see** the objects themselves, nor the **light** that passes through them or is reflected from them. Rather, it "receives" a very large number of electrical impulses, which come to it through the nerves.

From this flux of impulses reaching the brain, somehow, a picture of the world around us is built.

This picture is not built from the objects or the light rays that pass through or reflect from the objects around us.

It is not light that creates the picture that we see, but a flux of electrical impulses reaching our brain through the nervous system.

The picture that we "see" is not a picture of those electrical impulses, but a result of a complex "data processing" procedure, in which we have no control over the way it "takes place" nor do we understand its nature.

The process of creating an image, or the process of creating a thought is probably an automatic process that our brain creates without "consulting" us. Therefore, any change in this process, if that were possible, would change the face of reality as it "appears" to us.

---

18   For further reading, we recommend *The Brain: The Last Frontier* by Richard Restak (Appendix: Recommended Bibliography- editors).

This is somewhat similar to someone's voice reaching our ears through a phone receiver. If we were to suspect for a minute that the phone in our house was distorting (without our knowledge) the voice of an unfamiliar person who was calling us for the first time, would we be able to know that the voice we were hearing was the speaker's real, authentic voice?

Perhaps the voice coming out of the phone is not the speaker's voice, but a "restoration" of his voice.

The man speaking into the phone transmitter creates the sound waves that strike a light metal diaphragm and make it vibrate according to the frequency of the voice waves that are striking it. The vibrating diaphragm creates a series of electrical impulses that are sent through the phone lines. These electrical impulses may pass through thousands of miles until they reach the phone in our house. Here they cause another delicate diaphragm to vibrate in the receiver and to push the air that is next to the diaphragm into our ear, thus recreating the voice waves.

How similar is the voice that we hear on the phone to the speaker's real voice at the other end of the line?

If we have never heard his voice before, we will not be able to know for certain whether the voice we have just heard is similar or different to the "original" voice.

One could, through various means, send a flux of electrical impulses through the phone line, and in our phone, it might sound **as if** someone is talking to us, although no one is on the other end of the line.

In airplanes, an artificial "woman's" voice is heard, generated by a computer, which warns pilots when they are about to crash into another plane, during difficult conditions of visibility. This artificial voice is generated by and comes through a device termed a TCAS (**T**raffic **C**ollision **A**voidance **S**ystem), rather than being spoken by a human.

Special phones in the military and in the intelligence services "encode" the speaker's voice, so that someone listening to the conversation on an ordinary phone will only hear incomprehensible "garbled voices."

When we attach another device to the phone, which can "process" the electric signals reaching it according to some secret code, we will suddenly hear the speaker's voice clearly.

Which of the two voices is the real voice? The encoded one, or, the clear one?

In this case, we have no way of knowing for certain whether on the other end of the line there is a sensible person or a mixed-up creature, who cannot speak the human language. Perhaps both voices are equally real, the meaningless, "encoded" voice and the "clear" voice. They are both a reflection of a particular occurrence that we attribute to reality.

Perhaps direct access to reality is "blocked" to us through the system of signal reception and data processing that occurs in our brains.

Distortions of Reality

The first distortion of reality through **eyesight** occurs, when the light that falls on objects is reflected from them or passes through them partially, and not at all possible wavelengths. Similarly, voice, smell, taste and touch provide partial information of objects. They serve as "an intermediate medium" or "messengers," but are not the objects themselves.

> ➤ **The second distortion** occurs in the mechanism of **signal reception**, namely in the sensory organs themselves and in the "translation system" structure. This is where light, voice, smell, taste and touch are translated into a

series of electrical impulses, which flow from them to the brain, through the nervous system. The system of signal reception enables us to "sense" reality, but also distorts it, by transforming light in different wavelengths into a series of electrical impulses that are sent to our brain one after the other. The "translation" of the external signals into electrical impulses **slows** the rate of their reception, and thus causes another distortion of the signals.

> ➤ **The third distortion** occurs during the flow of impulses through the nervous system that slows, reduces and possibly distorts the signals once again.

Our brain does not receive the objects, nor does it receive light, color, voice, smell, taste, heat or cold, but a continuous flux of electrical impulses, which they themselves are probably not exact copies of the impulses that "left" the eye, the ear, the nose, the mouth and the hands on their way to the brain.

"The Triple Isolation"

> ➤ **The first disconnection** between us and reality is formed in the mechanism of information reception through the senses, and in the transforming of it into a flux of electrical impulses.

> ➤ **The second disconnection** is created in the nervous system during the transfer of information, especially due to the slow pace of its transfer. The flux of electrical impulses reaching our brain from the nerves is still meaningless, and only after a rather long data processing process (more than a tenth of a second), can we receive an "image," "smell," "voice," "touch" or "taste" through that same complex system that we call "the human brain."

> ➤ **The third disconnection** is caused during the data processing process in the brain itself, which creates another

distortion to reality, and probably also the deepest disconnection between us and that of the external reality.

Thus, our senses create our first disconnection from reality, the nervous system causes the second disconnection and our brain causes the third and deepest disconnection. We live in a system that is isolated from reality in a "triple isolation" and we insist on examining the world around us with this system.

This description of the system of signal reception and data processing is also conducted with the same system, which is isolated in the "triple isolation" from reality. Thus, it is also not a "true" description, but just one possible interpretation.

The "Perception" of Reality

The three disconnections from reality are probably automatic, over which we have no control. However, by growing up in a world with many external stimuli, this isolated system becomes the "reality itself" to us. We live and operate in a world created by our senses and our brain, and respond to it as if it were reality itself: each person with their own reality.

In which reality would we live, if, for instance, we would find a way to connect the nerves, which carry the information from the eyes to the "center of hearing" in the brain, instead of to the "center of vision," which usually processes these signals?

What would our reality look like if our nerves, which transmit the information from our ears to the center of hearing were instead connected to the center of vision?

It might well be that we would start "seeing" with our ears and "hearing" with our eyes.

The external reality would not have changed, but it would appear to us in a completely new and different way, not necessarily without beauty or meaning in this new, wider sense of the word.

Music, for example, would appear to us as images that change according to the rhythm, and with colors that greatly match the frequency of the voice waves and the melody. Its volume would appear to us as a bright or dim light, and in complete silence, it would be completely dark. Loud music would cause a dazzling brightness even in the darkest discotheque, and to prevent being blinded, we would have to cover our ears, since closing our eyes would not help in this case. All that our ears "see" would appear as sights, colors and changes of light and color, but it would be difficult to distinguish between defined shapes, as our ears do not have focus systems and they face in opposite directions.

Since our ears are on both sides of our head, they may not enable us to have three-dimensional vision or focus, or to spot objects from afar. On the other hand, "hearing" with our eyes may be a special experience, due to focused, three-dimensional hearing.

Even if we do not discuss this example any further, we can imagine what would happen if the smell and taste nerves were switched - we would probably "taste" the smells and "smell" the flavors.

After a while, we would have learned to "manage" and live in this strange reality, where our "view of the world" was completely different to the one we experience today, which seems to us as the reality itself.

That same external reality would look and sound completely different and the scientific equipment we would probably build, would be different, and built according to the structure of our "new senses."

Would this reality be "less real" than the reality we experience today?

The "new" reality would probably be just as real as the "old" reality. Both would not "really" touch the external reality, except through a human translation.

Is a book that is translated into one language less "real" than a book translated into another language?

Is a book we read with our eyes less "real" than a book translated to braille, which is read with the hands?

What would reality look like, if instead of switching hearing with vision, or taste with smell, we would find a way to increase the "reception rate" of these "systems"?

If we increase the ability of our eyes to see a wide range of colors, far beyond the range we "usually" see; so that we would be able to see beyond ultraviolet rays, X-rays and gamma rays in the short waves; or beyond infrared and the longest radio waves; or if we increase the ability of our eyes to receive even very small quantities of light, could reality look significantly and fundamentally different than the reality we know?

The darkest nights could appear to be brightly illuminated, just like at daytime, and we could see millions of "new" stars in the sky with a shining red light, stars that emit radio waves and stars that emit radiation in other wavelengths. Perhaps we would also discover other galaxies and additional stars in the galaxies we knew.

Perhaps some of the black holes - whose position could be identified through X-rays that, we assume, radiate around them, or through "Hawking radiation" - would cease being "black" and would also appear to have a color that we cannot name yet.

Perhaps, if we were to go out on a walk on a dark winter's day, we would not be able to be outside without the darkest sunglasses, to avoid being blinded.

While at home, we would be able to see the radio waves emerge through the walls, enter through the windows, through the ceilings and probably through the radio and television aerials. Perhaps we could learn about some of them. Maybe we could see radio waves while the radio was off, and as the announcer spoke, we could identify some of the information that reached us. Perhaps we could also discern the "colors" that were specific for that voice (our chances of identifying which radio waves were characteristic of what our favorite announcer looked like would probably be slim, due to the "excess information" received by our eyes).

If we manage to increase the rate of receiving information through eyesight and also significantly hasten the pace of the "translation work" that is done in the brain, will we be able to watch a movie without getting bored?

It is difficult to imagine a reasonable person who would sit in a dark hall for two hours, just to see static images change at times, with marginal differences between them.

The static images in a movie change at a rate of 24 images per second, as each image describes a static situation that is close to its predecessor. It is only the slow rate of signal reception and our brain's slow "translation" work that create the illusion of motion.

In fact, there is not one motion in the film! All the characters are "frozen" in each of the many singular images in it. It is just the slowness of the "vision system" which causes the illusion of motion on the screen.

What, then, is reality? Is it the one that moves continuously or the static one that changes every once in a while?

Perhaps both possibilities are only possible interpretations of some reality, both equally "good," but neither any truer than the other.

If we looked with our "rapid eyesight" at an old television

screen (built using a cathode ray tube), would we see the presenter of a program or any movie? No chance!

We would see a single spot of light on the screen, moving from place to place in a cyclical way, while changing its strength and color. There will be no trace of any characters or plot…

The "normal" slowness of eyesight and the swiftness of the moving bright light on the screen "enable" us to see figures in motion, although there is not even one character on the screen.

Our senses may create "disconnections with reality" that are innately human and may also be innate to other living creatures, but are these the only barriers that prevent us from direct access to reality?[19]

"Understanding"

There may be an additional barrier that is difficult for us to distinguish. This barrier is directly related to memory, and it is responsible for the sense of motion, for comparisons and for "understanding" reality.

This barrier is embedded so deeply in our consciousness that it becomes almost "transparent," and therefore disappears from our perception, leaving us with the feeling that we are able to sense and understand natural phenomena.

The concept of "**understanding**" creates the **illusion** that there is an external nature, one that is independent of us, which can be understood objectively by us through our senses and our brain.

Perhaps **understanding** is not a "revelation" of something that exists and is hidden from us, but one possible interpretation of

---

19   For further reading on these topics, we recommend reading *Meditations on First Philosophy* (Descartes), *The Critique of Pure Reason* (Immanuel Kant), *An Essay on the Immediate Data of Consciousness* (Bergson), *Science and Hypothesis* (Poincaré), *Between Science and Philosophy* (Leibowitz). Appendix: Recommended Bibliography (editors).

phenomena that are perceived in a "distorted" way by our senses, in the effort of creating order within them as we "understand" the term "order."

Perhaps "understanding" is the deepest barrier to becoming acquainted with reality.

Understanding "relies" on all that we have experienced, learned, read and tried out, namely, on the past and on memory, while reality is current, timeless, and to us – it may have only one property – "the generation of signals" that are perceived by our senses.

Examining signals in the place where they "meet" our senses may be the most promising way to expand our perception of what occurs outside of us, not as an image, sound, smell, or "concept" created by us, but as an external reality, regardless of whether we receive these signals or not.

What are those signals, which we can receive or process, and from which "the external reality" seemingly "emerges"?

Perhaps the signals that reach our brain through our senses and the nervous system are indeed the result of an external occurrence, but the exact nature of this occurrence may never be discovered due to the limitations of our thinking.

"Change of State" as a Key

Perhaps, in addition to learning about the limitations of human thought, we could also elaborate on learning about that "external reality" that "creates the signals for us."

The main way to open up possibilities beyond the ones we currently know about, may be by asking questions. This could be conducted by creating assumptions and then examining them.

Raising questions, even the most imaginary questions, would enable us to access phenomena that we could not have accessed until the question had been asked.

Perhaps it is not "the answer" or the "knowledge" or "the understanding" that would enable us to access the unknown, but the question.

Perhaps the answer to any question that we ask would seem to be reasonable to us, if the flux of signals reaching our brain and being translated into "reality" were to match our new assumptions "better" than it matched our previous assumptions that we perceived as "reality."

How can signals that affect our senses be created?

Could our senses observe a permanent, static thing, which has no motion or change?

Maybe the answer to the last question should be negative.

Researching motion could offer us a certain key to learn about an external occurrence.

How, then, is motion perceived by our senses?

We are probably not able to see or sense motion. We are probably only able to observe a change of state.

The change of state of an object or particle is the only thing we may be able to observe.

The slowness of our perception and the memory of a previous state create the illusion of motion or its "continuum," just like the example of the movie or the television.

The change of state is based on a comparison: Meaning the comparison of the new state to the previous state and the distinction between these two states.

This distinction can only occur through the memory that "freezes" one state and then presents it as compared to the new state, and thus reveals the difference between the two states.

The process can be presented mathematically as the

subtraction of one value from a previous, larger or smaller value:

When the result of the subtraction is different from "zero," there is motion.

When the result is equal to "zero," it is a static state.

Any state that appears in our brain and is different to zero represents motion, and any state that equals zero represents a static state where no change is taking place, it has no motion and for us, time does not pass in it.

Thus, the change is also an intellectual model, and not just an external occurrence, as it is built on memory and comparison.

The appearance of the concept of "change" in our thought, as a result of an external flux of signals reaching our brain, probably indicates a certain occurrence in the external world, as otherwise this concept would not have appeared in our consciousness.

In order for our nervous system to receive an external signal, this signal must be characterized by start and by end, namely: two states that are not identical.

Can we consider that - two states that are not identical - represent change or motion?

If we significantly "speed up" the rate of receiving and decoding the signals in our senses and our brains, will we still see things move?

The faster the rate of receiving and decoding signals, the "slower" reality will become according to our new perception, as possibly happens to flies, birds or other animals that respond faster than us to external signals. Reality would be "presented" to us as a series of images, which occasionally change, at a diminishing rate.

What would happen if we were to increase the rate in which we receive and decode signals, until our senses would receive a series of signals and instantly decode them? In this case, reality would probably "appear" to us as several static images. Would the images appear in a certain order that would match the time that passed?

If we accept that time is not a physical entity, and that it is just an "intellectual perception" rather than a direct observation of an external occurrence, we might also accept that the appearance of continuous occurrences depends only on the structure of our thought, and not on the direction time moves (or "the arrow of time") from the past to the future.

If the flux of fracticles reaching us from the universe around us is composed of singular fracticles, between which there is a certain distance, our view of the universe would change, as being not consecutive images, but a series of singular, discrete images.

Perhaps external realty, which does not depend on the rate in which we receive our signals, is not consecutive, but rather built from separate, disconnected segments.

Perhaps the connection between the phenomena and viewing motion, as a continuous thing, derives directly from the rate in which our senses and brain work, or put simply, just as the result of their slowness.

In order to elaborate more on this point, we will conduct an experiment with a swinging pendulum, and try to explore possibilities that we may have overlooked so far.

Can an experiment with a pendulum help to clarify the "secrets of motion"?

CHAPTER 6

The "Secrets" of Motion

"We must add to the old definition (which defined steady motion simply as one in which equal distances are traversed in equal times) the word "any," meaning by this, all equal intervals of time; for it may happen that the moving body will traverse equal distances during some equal intervals of time and yet the distances traversed during some small portion of these time-intervals may not be equal, even though the time-intervals be equal."

Galileo Galilei

A **pendulum** is a weight suspended from a pivot so that it can swing freely. If we move the weight to one side, lift it to a certain height above the floor and let it go, we will notice the nature of its course:

> ‣ The pendulum would travel downwards, until it reached its lowest point, from which it would continue rising up almost to the height from where we initially released it, at a point opposite to the starting point.

> ‣ At this point, the highest point at present, the pendulum would change its direction and move downwards, in the opposite direction to its initial direction of motion.

> ‣ The pendulum would swing downwards, to its lowest point, and then continue incessantly, and rise up again, almost to the point where it was at first, and thus one

"cycle of motion" would be completed.

In a real experiment, the pendulum would not "really" rise up to the original height, due to the losses occurred through the friction of the pendulum and the rope passing through the air, and the friction of the rope or the arm with the pivot at its connection point.

In order to learn about the "secrets of motion," we will overlook the losses of friction and focus on a detailed account of the "route" taken by the pendulum in each section of its movement.

What would we observe if we were to "enlarge" several sections in the route of the pendulum?

The Cycle of Motion

> At the highest point at which we released the pendulum, its speed as it would move towards the ground would be zero.

> When the pendulum would start moving downwards, it would "gain" speed and "lose" height. Thus, at its lowest point, the pendulum would reach its maximal speed, from which it would start rising upwards again, as it "loses speed," due to changing its height relatively to the state where it was at its lowest point.

> When the pendulum would reach its maximal height, it would change its direction, from left to right, for instance, and from going upwards to going downwards while increasing its speed, until it would start rising up again to the point where it was released, or very close to it. At this point, a new cycle of movement would start, and so forth.

Change of Direction

At the point of maximal height, the pendulum would change its direction, from ascending in one direction to descending towards the other direction, and from moving from right to left, to moving in the opposite direction, from left to right.

Do We Observe an Unusual Phenomenon?

This question is of great importance, since at this point, we see **a clear change** in the "behavior" of the pendulum, a change in its direction of movement, although, there seemingly is no significant change in the forces that "activate" it.

As mentioned earlier, "the cycle of motion" of the pendulum seems to be downward motion, from right to left, passing the lowest point, ascending from right to left, and at the highest point, changing direction, after which the pendulum starts moving again from left to right, descending downwards.

It is hard for us to observe the lowest point, because it is **not accompanied by a change of direction.**

At the lowest point, there is no clear change of direction, but one can notice **a gradual change** from descent to ascent and from acceleration to deceleration.

What will we observe at the highest point?

"The Highest Point"

What will we observe at the point where the pendulum stops its motion from left to right and still has **not started** its motion from right to left?

We will try to find out, in the most accurate way, what happens to the pendulum when it is **exactly** at its highest point.

In this state, the pendulum does not move from right to left and does not ascend, as it is already at its maximal height.

The pendulum does not descend, since it does not move from right to left.

The pendulum can only descend once it changes its direction and moves from right to left, since the arm's length from the weight up to the rotation axis is fixed.

Therefore, in what state is the pendulum, which is neither ascending nor descending, neither in a movement from left to right nor in a movement from right to left?

What happens at that "blink of an eye?"

What happens to the pendulum when there is no change in its position, in relation to the system it is in?

This pendulum is probably at rest, namely there is no motion in any of the directions in which the pendulum had moved so far.

If there is no motion, is it true to assume that time, which is the human measure of motion, also does not pass?

If there are two singular states in the motion of the pendulum – the peaks in each cycle, where there is a complete stop - could it be that at these extreme, "unusual" points, **motion does not take place and time does not pass?**

Perhaps we should also raise another question: are these two "peaks" in the course of the pendulum actually unusual?

How could it be that two states in continuous motion would be unusual, whilst no change has occurred in the forces operating the pendulum?

How did we get different results from equal causes?

Perhaps we should ask whether the two states that seemed unusual were not unusual at all, but it is only our **"slowness"** that does not allow us to observe that all other states are like them?

The Illusion of Continuous Motion

Perhaps these states are not unusual, and at each point along the course of the pendulum, there is **a complete stop**, when the pendulum is between two very close, successive states?

The factor that may prevent us from observing this phenomenon in the rest of the sections could be, "as usual," our slowness (our way of thinking, which creates **the illusion of continuous motion**).

Perhaps motion, just like time, (as a unit that describes motion) are both discrete or "quantal" and are not continuous as we tend to think, due to the limitations to the reception of signals and the rate of their processing in our mind.

Perhaps in this case too, the "limitations of human thought" create the illusion of motion.

Perhaps the pendulum stops at each point on its course, but we are unable to observe these stops.

Maybe there is no complete stop of time in the pauses of the pendulum, since there is still motion in the matter from which the pendulum is made. The molecules from which the pendulum is made move relative to each other according to the temperature of the pendulum; the atoms in the molecules move one relative to the other; and the electrons which surround the nucleus of the atom continue in their rapid motion around it (some 2,000 km per second).

Inside the atom nucleus, the protons, the neutrons, and all the particles that are known today keep moving, as well as, possibly, smaller particles that have not been discovered yet.

Do all these motions "really" go on continuously, or are they the same as the highest points of the pendulum?

The Quantal Nature of Motion and Time

- ➤ Could the quantal structure of motion be universal, with electrons, for   instance, obeying it during their rotation around the atom nucleus?

- ➤ Could motion be composed of **stops**, or "lack of motion," of segments where time **does not pass**, where there is no change in the position of the body in relation to the system in which it is?

- ➤ Could it be that time passes or flows just in the discrete segments of motion, namely between the stops?

- ➤ If there is a complete stop in the motion, how do bodies move and how do they change location?

- ➤ Do the bodies pass completely from place to place?

- ➤ Could bodies – all bodies in our world, disappear in one place, and reappear somewhere else, very close to where they have disappeared?

The ways in which our senses work, and all the equipment that we have created in order to enhance them, probably do not enable us to offer a clear answer to these questions.

The place where bodies "disappear" is the place where they do not cause a reaction to our senses or our equipment, and so we cannot know whether they are there or whether they have disappeared.

Perhaps to us, the cessation of motion is the duration of time that passes between the collision of two successive fracticles in the mechanism that receives signals by our senses and equipment.

Perhaps, when bodies reappear, we are able to observe their new position due to the "disturbance" they cause to the flux of fracticles coming from all around, from the universe that surrounds us.

Does the "disappearance" of bodies refer only to the large bodies, of such order of magnitude that they can be observed with the sense of sight, or does it also refer to smaller bodies, of the order of magnitude of atoms or smaller?

The Pulsating Universe

Since all the "large" bodies are made of atoms and sub-atomic particles, the "disappearance" of the atoms and sub-atomic particles would cause the disappearance of the large bodies, on condition that all atoms and sub-atomic particles would **disappear together**.

In this case, one may see the universe as a "pulsating universe," where all particles, and with them, all the small and large bodies, at once, disappear for a blink of an eye and reappear somewhere, very close to where they were before they have disappeared.

In such a case, there is no meaning as to how long bodies disappear, since time has no existence without motion. Since there is no body that moves, the universe can reappear after "forever." To us, that would be as if only a blink of an eye had passed, similar to the story of "The Sleeping Beauty."

If the "disappearance" of the atoms and the sub-atomic particles does not occur simultaneously, but rather each atom "disappears" independently, and each particle appears and disappears in its own time, then it would be very rare for a "large" body to completely disappear and reappear.

In these rare cases, this body might continue to appear and disappear, until the end of time.

It may be that since the universe has started, the number of "pulsating" bodies has increased, and today we may already be at the "pulsating universe" stage.

"Closing of Eyes"

If we are indeed in a pulsating universe, we cannot know it, since just like all other bodies, we also disappear and reappear alternately, together with all other bodies, which, creates the illusion that bodies do not disappear at all.

Perhaps this can be compared with a situation where two people look at each other, both close their eyes together, and reopen them exactly at the same moment, and so forth. Each one of them would then swear that the other always kept his eyes open.

We too momentarily blink our eyes without noticing it. When we close our eyes for a fraction of a second, our eyes are closed, and at that blink of an eye, we actually are not seeing anything. For us, during that split second, the world figuratively "disappears."

When we reopen our eyes, the universe reappears in a **slightly different** state than the state in which we "left" it previously.

We have grown so used to this automatic action of blinking our eyelids, that we do not notice that the universe in essence, "disappears and reappears."

We live as if there are no breaks in the flux of information reaching us through our eyesight, and possibly through our other senses as well.

The meaning of "a disappearing universe" to us is that, there is no external factor providing stimuli for the creation of electrical impulses by our senses, or that there is no external factor creating some reaction in the measuring equipment that we have constructed.

How can we know if the universe in which we live is a "pulsating" universe, where all stops occur simultaneously, or a universe where each atom is independent?

Pulsating or Continues?

> ➤ In a pulsating universe, there is no way to detect the complete stops, since the object that we intend to see or feel **disappears together with the** receiving or sensing **system**. This universe is like two people who play "the shutting eye game."

> ➤ In a universe that does not pulsate, or that only partly pulsates, one can distinguish between phenomena that occur during a complete stop, "in the background" of other phenomena that do not occur during a stop.

The difference between the phenomena would seem as if the phenomena, which occur when the body passes from place to place, before stopping or "disappearing," depend on the degree of disturbance created by the body to the flux of fracticles coming from all around.

These phenomena are not permanent, and depend on the rate at which the body changes position or, as usual: on its speed or acceleration.

In contrast to these phenomena, one could assume that phenomena that occur **only during stopping** would **always be permanent**, regardless of the motion of the body.

Can one distinguish between phenomena that occur during a complete stop, which do not depend on the speed of motion, and "regular" phenomena that occur in the change of place, and that are affected by the speed of motion or from the change of speed?

THE FABLE OF THE
TRAVELING CANNON

*"Thus it was characteristic of the special theory of relativity
that the concepts 'measuring rod' and 'clock' were subjected
to searching criticism in the light of experiment; it appeared
that these ordinary concepts involved the tacit assumption
that there exist (in principle, at least) signals that are
propagated with infinite velocity. When it became evident
that such signals were not to be found in nature, the
task of eliminating this tacit assumption from all logical
deduction was undertaken, with the result that a consistent
interpretation was found for facts which had seemed
irreconcilable."*

WERNER HEISENBERG

Let us imagine two cannons placed on trucks, each cannon on
"its own" truck.

Let us also imagine that the cannons occasionally fire a "training
shell" in the direction of travel of the truck.

What would be seen by an observer, who is at a monitoring
station to measure the speed of the shells as they pass in front of
him?

To make things easier, let us arbitrarily assume that the speed
of the shells is low and also that after firing, the shells move **in a
horizontal line and at a constant speed** (we will ignore the loss

of speed caused by the friction of the shell with the air and by its movement towards the ground).

Let us assume that Truck **A** stands in place with Cannon **A`**, and the cannon placed on it fires shells at a low (imaginary) speed of 1000 km per hour.

Truck **B** with cannon **B`** travels at a speed of 40 km per hour, and the cannon placed on it fires, **during travel**, shells whose speed is equal to the speed of the shells of cannon **A`**, namely: 1000 km per hour.

The observer at the monitoring station would see the shells of Cannon **B`** pass him at a speed of 1040 km per hour, which are 40 km per hour (truck) and another 1000 km per hour (cannon), adding up to 1040 km per hour.

What would our observer at the monitoring station see if Truck **A** would also start traveling whilst firing shells, and gradually increase its speed till it reached the speed of Truck **B** (40 km per hour)?

The observer at the monitoring station would probably observe that each shell of Truck **A** passes him at a slightly greater speed than its predecessor.

When Truck **A** reaches the speed of Truck **B**, the two shells will pass the monitoring station at the same speed, of 1040 km per hour.

The observer at the monitoring station would be able to predict what the speed of the next shell would be, if the driver of the truck told him (by radio) what the speed of the truck was at the time the shell was fired.

What would the speed of the shell be, if one of the trucks (truck **A**) stopped suddenly, and exactly at that same moment, the cannon would fire "the shell of the moment"?

What would happen if straight after the shell was fired from the cannon, the truck would continue on its way, as if nothing had happened?

After passing some distance, the truck would stop again and the cannon would fire another shell, after which the truck would again continue on its way.

In such a case, all the shells of the cannon on the truck **A** would pass the monitoring station at the same speed: **1000 km per hour**, regardless of the average speed of the truck.

Even if the truck were to travel very quickly, still all the shells fired by the cannon placed on it would continue to move at a constant speed of 1000 km per hour, as long as, the principle of firing at a complete rest were maintained.

When a certain action is conducted at a complete rest, there is no "addition" of speeds, since the speed of the truck always equals "0" at that moment.

The observer at the monitoring station knows about the average speed of the truck according to the moment it sets off on its way (the starting point), the moment it arrives (when it passes it at the end point), and the distance that has passed since it started traveling until it passes him.

The observer does not know what happens on the way.

The observer is unable to see the stops taken by the truck, as they are not in his field of vision, whilst the driver in the truck **A** notifies him of the speed of travel just one second before the shell is fired, or a second afterwards.

At the moment of fire itself, the truck is at complete rest and the scared driver does not notify anything.

Could such an unusual phenomenon exist in our world, which resembles the firing of the cannon on the truck?

Could there be a physical phenomenon that occurs at a complete rest, for which there is no "addition" of speeds?

Probably...

There probably is a phenomenon that does not obey the "Laws of Addition of Speeds," for which there is just one speed in a vacuum: the phenomenon of the speed of light.

THE SPEED OF LIGHT

"SAGR. But of what kind and how great must we consider this speed of light to be? Is it instantaneous or momentary or does it like other motions require time? Can we not decide this by experiment?

"SIMP. Everyday experience shows that the propagation of light is instantaneous; for when we see a piece of artillery fired, at great distance, the flash reaches our eyes without lapse of time; but the sound reaches the ear only after a noticeable interval.

"SAGR. Well, Simplicio, the only thing I am able to infer from this familiar bit of experience is that sound, in reaching our ear, travels more slowly than light; it does not inform me whether the coming of the light is instantaneous or whether, although extremely rapid, it still occupies time."

GALILEO GALILEI

Among the physical phenomena that we know, there is one unusual phenomenon, which is the speed in which light travels from place to place.

The speed of light remains constant (close to 299,792.458 km per second in a vacuum), when the source of light is at rest and even when the source of light is moving, in relation to the observer.

This result became clear following the famous experiment carried out by Albert A. Michelson and Edward W. Morley, which was first conducted in 1887, when Michelson was an army officer and a professor of physics, part of the academic staff at the Naval Academy at Annapolis.

Michelson and Morley compared the speed of light coming from a star moving in the direction of earth as it orbited around the sun, to the speed of light reaching earth from a perpendicular direction.

The speed of earth travel, as it orbits around the sun, is close to 30 km per second, a speed large enough to have covered any inaccuracy which could have occurred in Michelson and Morleys' experiment.

To their surprise, Michelson and Morley discovered that the speed of light reaching their measuring equipment is **one**, regardless of whether the source of light is moving towards the measuring equipment or moving perpendicular to the measuring equipment.

Stating that the speed of light (in a vacuum) is one (constant), posed a tough question to the physicists of the beginning of the twentieth century, and eventually led to the development and formulation of Einstein's Theory of Relativity.

In order to overcome the difficulty that arose from the revelation that light did not behave according to "the addition or subtraction of speeds" like in other phenomena, Einstein assumed that with the increase of body speed, there is a change in time and a change of body length in the direction of the movement.

If, in contrast to Einstein's assumption, we regard "time" not as a physical entity, but as an "intellectual model," and at the same time consider "distance" or the length of bodies to be intellectual models, then it will be difficult for us to accept that a change to our way of thinking can affect the behavior of bodies and phenomena that occur outside our brain.

Perhaps one should look for an explanation for the "strange behavior" of light elsewhere...

In this context, we may raise the question, could it be that light behaves like cannon shells which are fired during **a complete rest** over an average speed of a **moving** truck?

In such a case, could it be that light was created at a complete rest of the atoms, which is why its speed is constant?

If we provide positive answers to these questions, we may have to change the way we think about the subject of motion.

If indeed the "light photon" is created when the atom is completely at rest, which is an integral part of motion, its speed will remain the same, whether the source of light is in motion or at rest. Similarly, the light-receiving atom (the monitoring station) receives light at a complete rest when there is no motion.

The constant speed of light in a vacuum may point to how discontinuous motion is, or to its quantal structure, which may be seen as a basic phenomenon in our universe.

What is that thing, the speed of which we measure, which creates in our eyes and equipment the "translated" stimulus for the concept of "light"?

Perhaps when we discuss the speed of light, we mean the speed in which some "disturbance" in space is propagated.

Still, for the purpose of discussion, we can still refer to "particles" of light, the "photons," as some kind of material units.

The phenomenon of "lack of addition or subtraction of speeds" of light, points to the fact that we may live in a universe that does not "pulsate" together, and where each particle or group of particles "pulsate" independently.

The creation or reception of light at a complete rest can explain the constant speed of light in a vacuum, regardless of the motion of the source of light.

If indeed, the speed of light is constant **in a vacuum**, what about **"the slowing down"** of the velocity of light as it passes through transparent bodies? Should we see it as **a deceleration** or as another kind of change of speed?

We "know" that light travels at the high speed of approximately three hundred thousand kilometers per second, in a vacuum. However, its speed decreases slightly in the air, its speed decreases by about a third in water, and decreases further in glass. In a diamond, the speed of light is less than half of its speed in a vacuum, "only" about 124,000 km per second.

If we place several glass plates, one after the other, with an empty space between them, then a ray of light would move at a speed of 300,000 km per second up to the first glass plate. It would then "slow down" its speed inside the plate, and would leave it again at its initial speed, and it would continue with this pattern as it passed through the rest of the plates. After passing through the last plate, the ray of light would be moving at its original speed, namely 300,000 km per second (or more precisely: $C = 299{,}792.458$ km per second).

The fact that light "behaves" strangely in this way separates it from all other phenomena of which we have knowledge in the field of bodies and motion, and it raises essential questions regarding the phenomenon termed light, or the way in which we interpret what our eyes and equipment perceive.

How can it be that light slows down its speed in some types of mediums?

Even more amazing is the fact that light recovers its initial speed **immediately** after leaving the medium that "slowed down" its speed, "as if nothing had happened."

Does light accelerate as it leaves the glass plate?

Can we assume that light can pass, with no time, from a low speed to a high speed, in a kind of "infinite acceleration"?

Can there be an infinite acceleration?

What is the nature of a light ray that moves at a speed lower than the speed of light in a vacuum?

Can the light that travels through a diamond, for instance, still be a "light" when it moves at a speed that is less than half of its speed in a vacuum?

When a ray of light strikes a mirror, it is reflected at an angle **identical** to the angle at which it struck it. Namely, the light has not changed its speed due to striking the mirror's surface, otherwise, the ray of light that had changed its speed would have "created" a **different** reflecting angle from the "incident angle."

The reflection of light from a mirror may result from the "photons" colliding with the atoms from which the external layer of the mirror is made, and maybe also the layers next to it as well.

The light colliding with atoms of matter in the external layers causes **a change in the direction** of the ray of light but not a change in its speed.

How can light striking the atoms of matter of the internal layers, when light passes through the glass plate, cause it to decelerate?

When a ray of light leaves a medium such as a glass plate, from where does it receive the energy with which to increase its speed? Did the atoms in the external layers of matter "kick" the photon that passes nearby them?

Why does light that has "a long wavelength," such as a red light, move faster through glass than light that has "a short wavelength," such as a blue or violet light?

Can there be a simple and different explanation for the "strange" behavior of light that has so far eluded us due to our "way of thinking" or the interpretation we give to observable phenomena?

Can it be that a phenomenon that is measured by us, such

as measuring the speed of light, is built on "wrong" assumptions? Just like our observer in the monitoring station made assumptions about the speed of the truck, which were based on knowing when it started traveling and when it arrived, without knowing what had happened between these two points?

If speed is an intellectual model based on comparison rather than on external occurrences, we may examine two states, name one an "initial state" and the other an "end state," term the difference between them as "way" or "distance," and when we divide it into "time that had passed" between the two states, we term the result "speed."

We do not know what happens **between the two states** that we chose arbitrarily.

Perhaps we also refer to the speed of light in the same way, we assume that its speed decelerates when it passes through glass, according to the "angle of refraction" or according to the measurement of "the time it entered" the glass plate and "the time it exited" it (in fact, the measurements are more complex and indirect than this simplistic description).

What happens to a photon that moves between atoms of glass and occasionally collides with one atom or another that are in its way?

If the atoms of matter, between which is empty space, cause the photons to change direction, but do not affect their speed, as happens to the reflected light, the conclusion would be clear: light does not decelerate between atoms of matter, but passes **a longer distance**.

In order to demonstrate this point, let us imagine two twin rabbits, which both have the same running speed. We will release them near a forest of pine trees, where there is a nearby straight path. One rabbit will run on the straight path **next to** the forest, whilst its twin sister, for its own reasons, will prefer to run **in** the forest.

The "forest rabbit" would have to run in zigzags between the trees that prevent it from running in a straight line, while its sister would run in a straight, uninterrupted line (remember, both rabbits run at an equal speed).

The rabbit running on the path would pass a shorter distance, and would therefore reach the end of the forest first, whilst its twin sister, which, would run in zigzags, would still dawdle in the forest.

Perhaps the ray of light entering the glass, or any other mediums, would not decelerate but would pass a longer distance! The more atoms the light ray meets on its way, the more times it is diverted from travel in a straight line, the longer its route becomes, and we will interpret it as a slow speed.

Light which has a short wavelength would collide into more atoms on its way than a light which has a long wavelength, and it would therefore pass a longer way and arrive "late" to its target.

How would an increase in the temperature of the medium affect "the speed of light" in it?

When atoms in the medium are relatively close to each other, the "photon" would meet more atoms and have to pass a longer way between them, from its entry into the medium until its exit, or in the common terminology, "its speed is lower."

The increase in the medium's temperature which causes the atoms to move away from each other, would increase "the speed of light" (fewer zigzags), whilst the cooling down of the matter and the increase in its density would cause the speed of light to "slow down."

This phenomenon would be expressed with the "refractive index" of light in this medium, which expresses the relation between the speed of light in a vacuum and its speed in some mediums.

One can see that the speed of light remains constant in all conditions, and it is only our way of thinking that creates the illusion of different speeds in different materials.

A single "photon" can get "swallowed up" in the depths of an atom or have its linear motion diverted near an atom, but it **never changes its speed**.

Viewing the phenomena in this light strengthens Einstein's assumption about the speed of light as a constant value and eliminates the need to emphasize each time that it is just in a "vacuum."

Perhaps light is "born" whilst the atom is **at a complete rest,** and maybe it also gets "swallowed up" when it is at a complete rest. After going on its way, light moves at a **constant speed**, and it does not slow down or accelerate.

Perhaps the speed of light is the same both in a vacuum and in some mediums, since the space between the atoms must be mostly empty.

Perhaps what we term "photon" is a "cyclical disturbance" to the flux of fracticles, which is why the speed of light will equal the speed of the "disturbance," which equals the speed of the flux of fracticles.

In the same way, as the atoms divert the flux, they also divert "the disturbance," the number of fracticles at the same point, which is **different** from the number that was there "a blink of an eye" beforehand.

One may see light, or any other electromagnetic wave, as **a change** in the density of fracticles flux, whose speed equals the speed of the flux.

Where the disturbance to the flux of fracticles passes a longer way, between the atoms of the medium, it seems to us, from our point of view, **as if** there was a deceleration, just as we think of the rabbit that emerged last from the wood.

This poor rabbit, which ran among the trees, arrived last as it had passed **a longer distance** than its twin sister that ran on the straight path, who first reached the end of the wood and stopped to munch on some cabbage, which grew near the end of the forest.

To sum up this chapter, we can assume that:

> ‣ If the universe and all that is in it is composed of fracticles in motion, and if light represents to us a disturbance to the flux of fracticles, its speed will equal the speed of the flux of fracticles.

> ‣ If the speed of light is identical to the speed of the flux of fracticles, it must be the borderline speed in our universe, which is all composed of fracticles in motion.

> ‣ If the speed of light remains constant, both in a vacuum and in other mediums, perhaps other movements that seem to be slower than this speed are actually different than the way they seem to us.

> ‣ If there is just one speed in our universe, the speed of the flux of fracticles, all other speeds will seem different ("slower"), since the movement of bigger bodies is composed of the fracticles coming from different directions. Thus, the speeds of motion of other (bigger) bodies seem to be less than the speed of light.

Additionally, the speed of other bodies, which are also made of vortexes of fracticles, would seem to be solely as compared to the speed of light.

As observers, they would appear to us to have a different speed, since our senses only perceive **the change in the flux of fracticles.**

Perhaps the speed of light is not just the borderline speed in our universe, but is the only speed that exists in it.

Perhaps the only speed that exists in our universe is the speed of the flux of fracticles, and the phenomenon of constant speed of light is probably points to it.

CHAPTER 7

THE ATOM AND THE
SIZE OF FRACTICLES

"By convention sweet is sweet, by convention bitter is bitter,
by convention hot is hot, by convention cold is cold, by
convention color is color. But in reality there are atoms and
the void. That is, the objects of sense are supposed to be real
and it is customary to regard them as such, but in truth they
are not. Only the atoms and the void are real."

DEMOCRITUS

If we assume that all bodies are composed of smaller bodies, which, in turn, are composed of smaller bodies, we could assume that there is no limit to the small dimensions of the elementary particles.

Another option is that the structure of matter is continuous and not made of smaller parts, and thus it does not need a consolidating force to "hold it" in this state.

There may be other possibilities that our way of thinking (which deals with particles or continuous matter) conceals from us. That is because it is based on our daily experience with "large" bodies and on "basic software" that runs the way in which our senses are activated, the way in which data are received and processed and how the world around us is understood.

This basic software "tells us" what is "logical" and "illogical," based on our personal experience or on the experience of others, who preceded us and passed on their experience through written, electronic or other forms of information.

How can we know whether all fracticles in the universe have one size, if our way of thinking cannot "allow us" to choose between the different assumptions?[20]

Let us assume arbitrarily that the size of fracticles is determined according to the number of collisions that each particle experienced during its movement in the universe, from "the beginning of the universe" until today.

Perhaps the size of the fracticles depends on the "history of collisions" for each particle.

Perhaps a "head-on collision" between two particles would create a large number of fracticles, whilst a collision that is not "head-on" would create relatively larger but fewer fracticles.

In a universe with no attraction forces, there is nothing that would cause particles of matter to stay together. Therefore, with each collision, the particles would crumble into smaller particles, which would move along new courses according to the direction of movement of the colliding particles and their size before collision.

The particle "fragments" would keep getting smaller as they collide into other particles, until reaching the size where we term them "**fracticles**," in order to distinguish them from the "larger" particles that are currently known (electrons, protons, neutrons, etc.).

The smaller the fracticles get, the slimmer their chances of colliding with other fracticles, whilst fracticles that are larger would continue to collide and get smaller, and finally, they might all reach a more or less uniform size.

Perhaps most of the fracticles have already reached such a small size that they can barely get any smaller, and we can consider it as the final size of the fracticles at this stage of the "development" or "history" of the universe.

---

20   See Chapter 5, "Limitations of Human Thought."

We may be able to talk about fracticles as bodies of a uniform size, at least statistically, since the frequency of their mutual collisions decreases, the smaller they get and the larger the "expanding universe" becomes, as is commonly thought.[21]

If fracticles were bigger, their "collisions" with the atoms would have caused them to leap in zigzags, just like small particles that float in liquid or gas "leap" when the liquid or gas molecules hit them, or just, as an analogy, like a rubber ball floating in the water when a dolphin hits it with its nose.

The movements of the atoms, electrons, protons, neutrons, and other elementary particles may indicate the size of the fracticles, since the movement of the various particles may be caused by the fracticles.

Is there any phenomenon that could positively or negatively indicate the size of fracticles?

In order to answer this question, let us imagine a "dolphin show" in a round pool, in which a rubber ball floats on the quiet water.

When a dolphin emerges from the bottom of the pool and hits the ball with his nose, the ball "leaps" into the air, and falls back in the water at a point that is far from where the hit occurred.

As the ball falls back into the water, another dolphin emerges from another direction and hits the poor ball again. The ball then "leaps" in a new direction, different from the previous direction.

Each time the ball falls back down into the water, it hits the water and it leaps again into the air, to the delight of the audience.

If dolphins were transparent, we would only see the ball bouncing on the water with sharp movements, to different directions, with no explanation of this strange, irregular behavior.

How would the ball behave if there were small fish in the

---

21   See Chapter 1: "Why Do Apples Fall?" and Chapter 8: "Background Radiation and the Big Bang."

pool (instead of big dolphins), which would swim very fast, in straight lines, and would collide with the ball each time it was in their path?

In spite of the many collisions, the ball would "display" seemingly calm behavior, apart from a slight tremor or very small movements when the collisions it would get from the fish would not be evenly delivered from all directions.

We might notice that the ball moves in a certain direction, while most of the fish move in that same direction. Furthermore, the number of hits the ball would get from this direction would be higher than the number of hits it would get from the opposite direction.

How would a small particle floating in the water behave when the small water molecules, which are in constant motion, would hit it?

If we refer to the small particle as we refer to a rubber ball, and refer to the molecules, which are close to it in size, as dolphins, then we would expect the particle to move about in zigzags, as it would get unevenly hit by the water molecules.

In 1827, the British botanist Robert Brown observed and researched the irregular motion of small particles that float in a liquid or gas. This motion was named after him – "The **Brownian Motion.**"

This constant, irregular motion of small particles, floating in a liquid or gas, probably derives from collisions with the liquid or gas molecules.

The liquid molecules, which are in motion, are close in size to the size of the floating particles, and thus their effect on the latter is evident.

The particles can be particles of stamens that float on the water or particles of smoke in the air, or any other particle whose size is not bigger than about one micron (one-thousandth of a millimeter).

If the molecules were several orders of magnitude smaller, then a small particle floating on the water would get a more or less equal number of light hits from all directions, and would stay in its place. Alternatively, it would move towards where it had received the smallest number of hits. In such a case, we would not observe the "Brownian Motion" in its known format.

The size of the molecules in relation to the size of the floating particles is directly "responsible" for the "Brownian Motion," that strange, irregular motion of small particles in a liquid or gas.

What causes the motion of the molecules themselves, and what moves the atoms?

Just as the bounces of the rubber ball hint at the size of the dolphins, so does its relatively quiet motion in the pool, indicate the minuscule size of the fish.

Since we did not observe an irregular motion of atoms or molecules such as that of Brownian motion, we may perhaps assume that the fracticles that hit them are several orders of magnitude smaller than the molecules, the atoms or any of the different "elementary components" of the atom.

The constant motion of atoms and molecules may be the direct result of their collision with various numbers of fracticles coming from opposite directions.

Until the days of Louis Pasteur, no one had suspected that small, single cell creatures would be the main cause of the many diseases and plagues that caused the deaths of many people in Europe in the nineteenth century.

During that period, scientists, doctors and clerics, all gave completely different explanations for those phenomena and did not take suitable precautions to prevent infections, since they could not see the bacteria without a microscope and they believed in other, sometimes mystical causes.

Are we entitled to reject the existence of fracticles because there is not and there may not be a way to see them?

The "Birth" of the Atom

If a small rubber ball is placed in a pool with **transparent fish**, swimming fast in all directions, and in the center of which there is a stationary large rubber ball (which gets more or less equal fish-hits from all directions), the small ball would get fewer hits from the direction in which the large ball is located. This is because the large ball would serve as a "barrier" to the motion of the fish, and therefore, the small ball would move towards the large ball.

If the small ball has initial speed that is not in the direction of the large ball, or is moving away from it, it would then move towards the large ball, whilst deviating towards the initial direction of motion (alongside a curved line where the large ball is on the perpendicular to the tangent of that curved line).

The "Shadow Triangle" (fewer fish) will always be between the large ball and the small ball while the "Repulsion Triangle"[22] (more fish) is beyond the small ball on the line adjoining the centers of both balls, so that the small ball will always be at the vertex of both triangles.

If the direction of the initial motion of the small ball would be towards the large ball, it would "fall" and cling to it in the end.

If the direction of the initial motion of the small ball would be away from the large ball, it would move away from it as long as its speed were high enough, or else it would go back and "fall" towards the large ball if its speed were lower than a certain critical value.

---

22   See Chapter 2: "Distance and Repulsion."

Regarding all other directions, if the speed were sufficient, then the small ball would enter a circular or elliptical motion around the large ball. That is as long as it received more hits from the outside (from the direction of the pool's rim), and an equal number of hits from all directions, but fewer hits from the center (from the direction of the large ball).

In such a case, the small ball would have to continue to move in a circle or in a shape that is similar to a circle around the large ball, as long as equilibrium was maintained in the system.

To us, as observers of the spectacle, where we could not see the fish (which, as mentioned earlier, are completely transparent), it would seem as if the large ball "pulls" the small ball towards it, and causes the small ball to move around it in circles.

In order for the small ball to move around the large ball in circles, only two conditions are needed:

> The reduction of the "flux" of fish from the direction of the large ball.

> The small ball should have initial motion, and at a sufficient speed.

The direction of this motion could be in any direction, as long as it is not towards the center of the large ball or from its center and onwards.

In the "Fable of the Pool," the large ball can represent the "atom nucleus" and the small ball can represent one of the electrons that revolve in orbit around it.

It is likely that the electron is not like a rubber ball, or like any other body that we know of in our daily lives, and that the "Fable of the Pool" merely illustrates a basic explanation of the idea.

If we replace the fish with "fracticles," the large ball with "a large cluster of vortexes" (also made of fracticles) and the small ball with "a small cluster of vortexes" (made of… fracticles), we would "witness" what could be described as "the birth of the atom."

In the following pages, there is a graphic description to demonstrate this idea:

Illustrations No. 18 and No. 19 describe the effect fracticles have over small bodies and the directions of speed in any consequent points, as the "Repulsion Cone" and the "Shadow Cone" (instead of the "Repulsion Triangle" and the "Shadow Triangle") cause a change in the direction of motion at these points.

In Illustration No. 18, the motion of the small ball is parallel to the page plane and in Illustration No. 19 the plane of motion is tilted, in relation to the page plane.

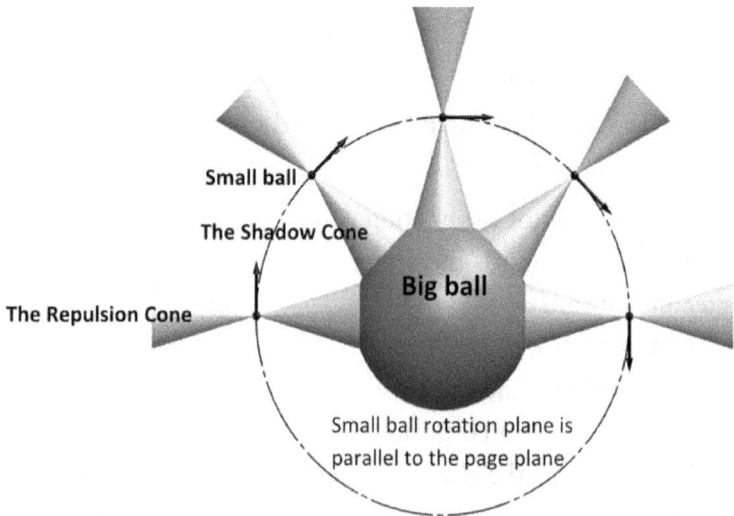

Illustration No. 18

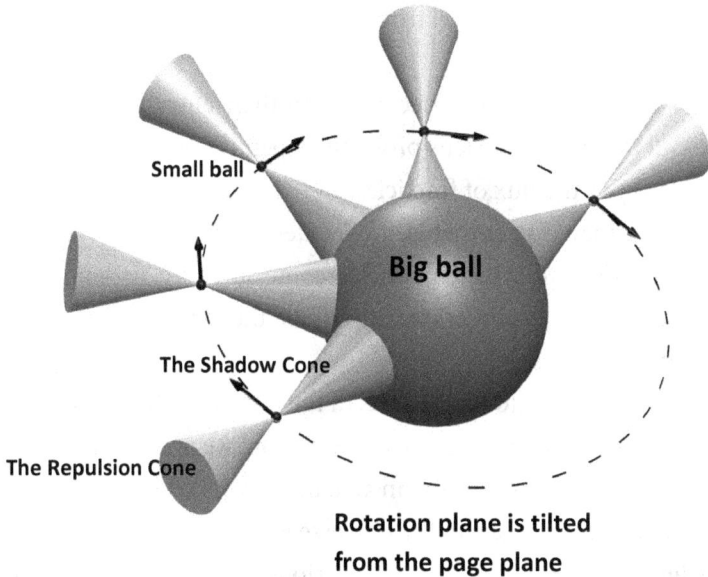

Illustration No. 19

This explanation of the structure of the atom is also applicable to systems like the solar system with the planets rotating around the sun, or to systems of other stars and planets.

The description of the fish and the rubber balls in the pool indicates that in a universe in which "numerous" minuscule fracticles move about in straight lines, any disturbance to their motion would cause a focal point, " vortex" or "cluster of vortexes," around which other "bodies" or small "disturbances" would revolve.

The large "disturbance" (the large rubber ball) would be what we term the "atom nucleus" and the "small disturbances" (the small rubber balls) moving around it would be the electrons.

Any additional "small disturbance" or "small disturbances" around the rather large "initial disturbance," would lead to the creation of additional "electrons" which would move around the central "nucleus."

The creation of the electrons and their entry into an orbit around the atomic nucleus probably depends on a certain lack of symmetry in the flux of fracticles.

One may expect that at the "center of the universe," if such a place exists, to which fracticles come in more or less equal numbers from all directions, atoms of the structure we know would not be created.

If we manage to locate areas in the universe where atoms are significantly different to those that we know, we may then observe the center of the universe or an area that is very close to it.

In any other place in the universe where the flux of fracticles is slightly or very different in the various directions, the creation of the atoms can be described according to several steps:

Step One

The creation of very small, spatial, fracticle vortexes, which "disturb" the free motion of fracticles, moving in a straight lines.

Step Two

The creation of "atomic nuclei," composed of vortex clusters or "clusters of disturbances," which are very close to each other.

Step Three

The creation of "small disturbances," that are far from the nucleus, and which move in a different direction than towards the atomic nucleus or away from it.

Step Four

The entry of those small disturbances, into orbits, around the atom nucleus.

This description of the creation of atoms is essential for any location in the universe that is somewhat distant from its center, where fracticles coming from all directions do not have exactly the same density.

This slight asymmetry is enough to create vortexes, which become "disturbances" that consolidate the atomic nuclei and electrons that rotate in orbits around them.

Since there is "no limit" to the continuation of the process, one may say that the process of creating atoms out of the undisturbed flux of fracticles coming from every direction has "all the necessary time in the world."

The Big Bang Theory enables this process to occur in a very short period of time according to the same lines of development, anywhere, apart from the very center of that "big bang," if it did indeed occur.

The Big Bang Theory is possible, but it is not essential to the creation of atoms in their familiar format. Perhaps it should be changed, or even forsaken.

What "Really" Exists?

"Beyond bodies and space there is nothing which by mental apprehension or on its analogy we can conceive to exist. When we speak of bodies and space, both are regarded as wholes or separate things, not as the properties or accidents of separate things."

EPICURUS

"We cannot solve our problems with the same thinking we used when we created them."

ALBERT EINSTEIN

The universe, according to a description of the atom as a "cluster" of "disturbances" in a vast flux of fracticles coming almost equally from all directions, **contains nothing else**.

The mutual influence between the elementary particles of the atoms and between the atom, and its close and distant atomic neighbors, occurs, only through the collisions they get with the fracticles coming from around them.

The concept "force" has no meaning, apart from the description of the hits that fracticles give to other fracticles (organized in "vortexes" or "disturbances" or "atoms" and molecules).

Perhaps the change in the flux of fracticles is the only thing that can be distinguished by our senses or by the equipment we build.

When the flux of fracticles is undisturbed, it is "transparent" to us and we have no way of discovering it or identifying it.

When the flux of fracticles is disturbed, concentrated points or centers of "reality" are created. We can detect them and term them "matter" or "energy" or any other way chosen by our brain and the language we created to present things to us.

Where there is **no change** in the flux of fracticles, there is **nothing** for us.

Do atoms, electrons, protons, neutrons and other particles "really" exist?

We have no clear answer to these questions, and we probably will not be able to know what an atom "really" looks like, since it is too small for us to see it directly with light.

In our thoughts, we build an imaginary model of the atomic structure and then we compare the results of the experiments and the observations we conduct, to see how good the match is.

If it is a good match, we tend to accept the (imaginary) model and refer to it as "real," until other results or observations show up that it does not suit "our" model, or which we cannot resolve with this model, according to the "plan" called "logic."

In such a case, we must change the model or replace it with another, which would better suit our observations.

None of the models are the "actual truth": they are all **imaginary models**, created by man, rather than being anything in reality.

It does not matter at all if a certain model is accepted or not accepted by scientists, all models are just an "aid" to the way in which we interpret the world around us.

The universe may exist, but we have no way of seeing it.

Let us return to the atomic nucleus, which, as is commonly assumed, contains different particles (protons, neutrons, etc.) held together by the "strong force" and the "weak force."

What are those forces that are able to hold together particles charged with an identical electrical charge (which is commonly assumed to be positive) which "strive" to get away from each other?

What can be the reason that bodies charged with an identical electrical charge would "strive" to stay away from each other?

Alternatively, what is the thing that "tries" to distance common electrically charged bodies from each other?

If there are two balls in a pool of water, which are very close to each other, fish could not penetrate between them and distance one ball from the other, while the fish from the outside hit them incessantly, pushing them together.

Even the slightest distancing of the two balls from each other would enable a few fish to "infiltrate" between them and trigger "repulsion" that would distance them a bit from each other and allow additional fish to "infiltrate" inside.

Under similar conditions, the "force" that pushes relatively "large" bodies close to each other could be very large when they are very close to each other, in comparison to the size of small bodies that "try" to push them closer, or to distance them from each other after they have slightly moved away from each other.

Once the distance between the large bodies grows a bit, "the force which pushes them together" will decrease significantly and behave according to the "Repulsion Cone," as it "behaves" in relation to gravity.

If an elementary particle (a neutron) infiltrates into the "package" of the atomic nucleus, it can distance the vortexes (the elementary particles) and allow an infiltration of additional fracticles into the package.

The infiltration of fracticles will trigger strong repulsion forces between the particles (fracticle vortexes), which would eventually cause the package to be dismantled and the particles to be "violently" pushed out of the boundary of the atomic nucleus.

Neutrons that would be ejected in such a way would infiltrate into other atomic nuclei and cause in these atomic nuclei a **nuclear fission**, while emitting additional neutrons, and thus they would cause a "chain reaction."

The chain reaction can only take place if there are enough atoms into which the neutrons can infiltrate before they leave the area where the "fissile material" is located.

The minimal amount of fissile material that enables a chain reaction to develop, like a snow avalanche, is called a "critical mass."

What enables a neutron to infiltrate between the vortexes in the atomic nucleus?

Perhaps the neutron does not have the shape of a vortex and it does not revolve in space around any axis. It has small dimensions and the fact that it does not revolve enables it to infiltrate between the vortexes without any rotational motion (its own self rotation) colliding with another rotational motion (the vortex motion of the fracticles in the proton).

It is commonly thought that the neutron has no electric charge (hence its name), which is why it is not repulsed by protons with a positive electrical charge.

Protons and other elementary particles, which have grown apart from each other and enabled the fracticles to infiltrate between them will keep growing apart at an increasing speed, and will cause the atoms that they collide with to move in the same direction, though at a reduced speed, due to their relatively large dimensions.

This motion of atoms in all directions very rapidly increases the volume that the critical mass has in space and creates what is commonly termed "an atomic explosion."

The description of the atom presented here (including the atomic explosion and the way it occurs) is not meant to replace the commonly accepted theories that "successfully" led to the creation of the atomic bomb, and atomic reactors. They are meant to point to a new way of understanding this concept, without requiring the complex and largely "artificial" structure of what is termed today as "the standard model of the atom."[23]

The multiplicity of the elementary particles and the complex structure that is currently attributed to the atom, make this structure "suspicious." Perhaps a new, simpler structure will replace it soon.

---

23 In these contexts, see **Appendices – "Glossary"**: The Standard Model of the Atom, Energy, Fissionable Matter, Critical Mass, Chain Reaction, Atomic Bomb. In the **"List of Notable People,"** see Einstein, Bohr, Gamow, Rutherford, Heisenberg, Oppenheimer, and Feynman (editors).

CHAPTER 8

- ➤ Black Holes
- ➤ Background Radiation and the Big Bang
- ➤ Finite, or Infinite?

BLACK HOLES

"The astrophysical roles of black holes are by now widely appreciated. Their importance for understanding quantum gravity is also accepted. Still, most would regard the black hole as a curiosity devoid of relevance for everyday physics. After all, astrophysics is far from the laboratory, and quantum gravity may never be testable experimentally!"

JACOB BEKENSTEIN

"A black hole has a boundary, called the event horizon. It is where gravity is just strong enough to drag light back, and prevent it escaping. Because nothing can travel faster than light, everything else will get dragged back also."

STEPHEN HAWKING

Could it be that even bodies that we cannot see and whose existence we only assume, based on theoretical assumptions and indirect observations, can still be described when we assume the existence of fracticles and motion?

In the 18th century, the English scientist John Mitchell and the mathematician–astronomer-physicist Pierre-Simon, marquis de Laplace, first devised the idea of super-dense stars that underwent "a complete collapse of gravity."

In 1969, Prof. John Archibald Wheeler, the theoretical physicist, also used this definition, when he called these astronomical objects "black holes." This term stuck, and they are still known by it today.[24]

One of the candidates to be described as a "black hole" is the main star of the double (or triple) star system **V404 Cygni** (or it is a micro-quasar), which is about 5000 lights years away from us, and is in the Cygnus constellation.

Can one describe bodies such as black holes through fracticles?

"The Creation of a Black Hole"

Let us assume that as a result of an uneven dispersing of minuscule vortexes in space (such as in the case of a collapsing star), regions would be created where the vortexes density is significantly higher than their density in other regions. This could create an area or region that would be nearly **impenetrable** to fracticles.

If the flux of fracticles reaching this region from all directions cannot pass through it, the number of fracticles "getting caught" in the region will increase. Similarly, the number of vortexes will also grow and their density will be increased.

Fracticles leaving the vortexes and moving **in the dense region** in linear motion going outward will probably collide into other vortexes, and their motion will be interrupted or stopped.

Only fracticles that are very close to the surface of the **event horizon** of that "dense region" or "black hole" may manage to leave it.

These fracticles coming from beyond the "event horizon," if perceived by our senses with the aid of the scientific equipment at our disposal, may seem to us like very weak radiation around the black hole, in the background of other, much stronger radiation, coming from beyond them, from the rest of the masses in the universe.[25]

---

24 Quote: Prof. John Archibald Wheeler, (editors).
25 On the 22nd of July, 1999, NASA launched Space Shuttle Columbia into space,

One can notice a shortage of fracticles or "a shortage of information" only when compared with other information, or in other words, the black hole creates a certain "shortage" in the "normal" flux of fracticles, and this can be noticed when we compare the flux of fracticles coming from all directions apart from where the "black hole" is. For us, the "shortage of information" would appear in this direction as "a black hole" or as "super-dense mass," due to the high density of the vortexes.

If a "black hole" is discovered with any certainty, it could provide some validity to the existence of the flux of fracticles.

If a material body approached a black hole, it would keep "absorbing" the flux of fracticles from all directions, apart from the direction in which the black hole is. If it does not "escape" in time, it will fall and "disappear" inside it, while somewhat increasing the density of that "black hole."

The fracticles from which that body is composed would join those that are already in that region, and this will increase the disturbance to the passage of additional fracticles, and the "black hole" would become blacker.

The Fate of the Black Hole

Do all "black holes" in the universe have one fate?

What would be the fate of a black hole that is close to the "center of the universe," if there is such a center, and what would happen to a black hole that is somewhere closer to the "edges" of the universe, if there are such edges?

Perhaps a black hole that is somewhere on the edges of the universe does not "receive" an equal "rain of fracticles" from all

---

under the command of **Eileen M. Collins**. Its main mission was to install an advanced astrophysical telescope named **Chandra** (after the Indian physicist) to research **X-rays** from space. Such rays probably derive from hot gaseous areas of exploding stars, such as quasars, and according to current assumptions, this is also the weak radiation that surrounds black holes (editors).

sides, so that the number of fracticles reaching it from the center of the universe is much higher than from other directions. Thus, such a black hole would start spinning faster and faster around some axis, according to the difference in the number of "hits" it got from different directions.

When the rotational speed of that "black hole" would pass a certain borderline value, fracticles and vortexes would start "leaving" its perimeter and move in straight lines in all directions.

This effect can appear to an outside observer as if this area in the universe has started to shine and expand, and where the black hole once was, there may be a "new star," called **Nova**.

Such stars are rarely observed, as they are at very great distances from us, and they may be proof of the collapse of a "black hole" closer to the "edges" of the universe.

A black hole that is in the "center of the universe" or relatively close to it would "absorb" large quantities of fracticles, more or less in equal numbers from all directions.

Such a black hole would not have a "self-regulation mechanism," and so its density would grow until just a limited and dwindling number of fracticles would manage to pass through it from one side to the other.

Furthermore, such a black hole would become blacker and blacker, as it would not let "information" pass from one side to the other, or to leave it.

When the number of collisions of the dense fracticles **inside** the black hole would be greater than the number of collisions with fracticles coming from the rest of the universe, the black hole would start expanding very rapidly, due to the violation of the equilibrium.

Such an expansion may turn the black hole into a **Supernova**, which is observable even from a very great distance.

Supernovas were observed in the years 1006, 1054, 1604, and 1885. The last one was observed on February 23, 1987, in the Large Magellan Cloud, at a distance of one hundred and seventy thousand light-years away from us.[26]

If we were to manage, or to detect a direction in space, where more "supernovas" are detectable, than in other directions, we might be looking towards the "center of the universe," and perhaps we could also know our location in relation to this speculated "center."

The description of black holes may enable us to examine a few other options that lie within the idea of a flux of fracticles moving in a linear motion at the speed of light.

Where is the "Missing Matter" or the "Dark Matter?"

Perhaps we should look at the universe as one unit where "vortexes" or "masses" are concentrated in clusters called stars, galaxies, black holes, or nebulas, between which is **a massive flux** of fracticles moving in straight lines at light speed in all directions, in which there are many small vortexes which have not yet consolidated into "glowing masses."

This way of looking at things could lead us to think that most of the mass in the universe is not where we would expect to find it, i.e. in masses which themselves are visible (stars), but **in the space between them**.

About 60 years ago, the astronomer **Fritz Zwicky**[27] announced that most of the matter in the universe is "missing."

More precise observations that were conducted during the past decade, and also those that examined the rotational speed of

---

26   For further reading, we suggest reading *National Geographic*, issue No. 17 (October 1999), pages 43-45, and in the maps of "The Milky Way" and The Universe" which are attached (The editors).

27   For further reading, we recommend looking at the appendices "**List of Notable People**," "Zwicky, Fritz" and in the **List of Concepts**: "Missing Mass" (The author).

galaxies around themselves and the way in which they move in the universe, show that most of the matter in the universe is not visible.

Where, then, is the "missing matter" or the "dark matter" of the universe?

Perhaps the place where scientists look for this "missing mass" is not the place to which the telescopes, radio telescopes or the most sensitive CCD cameras are pointed.

Perhaps the "missing mass" or the "dark matter," consists of the "small vortexes," which disturb that flux of fracticles between the stars and galaxies.

These large numbers of small vortexes probably contain what we term the "missing mass" or "dark matter."

Background Radiation and the Big Bang

"Philotheus: It is not that it may see the infinite, has no sense from which you are claiming this conclusion; because the infinite cannot be the object of sense; and though those who ask to know this because of the sense, is similar to the one that would see with the eyes, the Substance and essence; and those who deny this thing, because it is not sensitive or visible, can be to deny its Substance and being".
Bruno Giordano

Astronomical observations in faraway galaxies and in galaxy clusters, point to the fact that galaxies and clusters are getting further away from us and from each other. This phenomenon probably indicates that the universe is expanding.

According to the redshift of the light of galaxies that reaches us, the calculated rate at which they are moving further away is greater for distant galaxies than for closer ones. This phenomenon indicates that not only is the universe expanding, but it is expanding **at an accelerated rate**.

If, indeed, the universe is expanding, its temperature is probably dropping, since its average density is getting smaller.

If the universe is getting cooler, it probably used to be much hotter, and if its density is decreasing – then its density in the past was probably much greater.

Scientists tried to calculate and find what the temperature of the universe could have been "the moment it was started," when it was all concentrated in a minuscule volume, from which it started expanding at a huge rate, and cooling down.

Those who support the Big Bang Theory claim that the universe started in one point in time, or at the beginning of time, and at a most minuscule volume, somewhere in space, if there was an empty space before that...

Since the initial temperature was very high and the rate of expansion was very fast, this "phenomenon" was termed "The Big Bang."

The term was given by Fred Hoyle, who did not believe in the Big Bang Theory.

According to some calculations, the Big Bang could have taken place approximately 12-15 billion years ago[28] and could have looked like a vast outburst of energy, from which all masses in the universe were later created.

Many see the Big Bang as "the beginning of time" and "the start of space or the start of the universe."

People who support this theory think that the "death of the universe" may occur when the Big Bang energy subsides. The universe will stop expanding and the "forces of gravity" would force it to collapse with increasing speed and temperature, towards a violent explosion termed "The Big Crunch."

A "short time" after the Big Bang, a few hundreds of thousands of years later, the temperature of the universe dropped, from an imaginary temperature of 10^{32} degrees Celsius (or more) to a temperature of just about 3000 degrees Celsius.

---

28 New calculations based both on measurements taken by the Hubble Space Telescope and on up-to-date measurements taken with radio-telescope equipment regarding the rates, at which faraway galaxies are getting further away, show that the universe may be up to 12 billion years old, and it may even be "younger," and "only" be 10 billion years old. It was also claimed that the rate of expansion of the universe is approximately 21 km per second, per million light years." (*Popular Science*, October 1999, p. 21) (editors).

At this temperature, characteristic radiation was emitted, which today should mostly be found in the field of "micro waves" and which, according to scientists, should be reaching us today.

Due to the expansion of the universe, this radiation, which is assumed to be remains of the Big Bang, would be seen as radiation emitted from a "black body" and should be today less than 3 degrees above the "absolute zero."

As early as 1948, **George Gamow**, together with **Ralph Alpher,** predicted the existence of weak radiation derived from the beginning of the universe, when it was much hotter than it is nowadays.

In 1956, **Arno Penzias** and **Robert Wilson** discovered weak radiation in the wavelength range of "microwaves," which comes to us from all directions and almost with the same wavelength and intensity. It has since been termed "cosmic background radiation."

Many scientists saw this radiation as a confirmation of the Big Bang Theory.

Does the background radiation indeed confirm the Big Bang Theory?

Could radiation arriving to Earth from all directions have witnessed a special event that occurred at a certain point (perhaps **the first point**) in space and in time?

In this chapter, we will not address those who claim that it was not a single point but the whole universe, and therefore there is no single starting point, but rather all the points in the universe are the beginning. In this chapter, we will also not address the theory of cosmic inflation, which was probably developed in order to justify the average homogeneousness of the universe.

In previous chapters, we saw that space and time are creations of man, rather than properties that were part of an external reality, and thus…

If the Big Bang did indeed occur in a certain place, and from there matter spread more or less equally in all directions, one can view this "phenomenon" as a hot ball, which is growing at an incredible velocity.

It is hard to assume that we, Earth, the Solar System and the Milky Way galaxy, where the solar system is located, are at the same point where the Big Bang occurred, namely at the current "center of the universe."

George Gamow's prediction from 1948 and the discovery of background radiation in 1965 thrilled many scientists and astrophysicists, who saw them as almost complete evidence of the Big Bang Theory.

Was their joy not a bit premature?

Perhaps instead, we should be somewhat doubtful of such a nice match of "reality versus prediction," and we should present a number of questions:

> If the radius of the universe, as it is observed today, is 12-15 billion light years, and we are not at its center, why should we not receive the cosmic background radiation **most prominently** from the direction where the start of The Big Bang occurred?

> Perhaps we may assume that the region where that radiation was created should be "rich" with "microwaves" and perhaps also with radiation in other wavelengths. This would suit different temperatures that were emitted before and after the "radiations" from that "hot ball."

> Perhaps we may also assume that the other directions in space where the Big Bang did not occur should be "**dark**" and without any background radiation.

> If the Big Bang occurred in a certain place, in which we are not located today, the remains of that "bang" should

reach us especially from the direction in which it occurred.

> If the masses created at and after the Big Bang (between
> 100 thousand to a billion years) move further away from
> the center of that "bang" at a speed that is significantly
> lower than the speed of light, one can assume that radia-
> tion derived from the Big Bang and moving at the speed
> of light **has passed us long ago.** In addition, it is continu-
> ing to get further away from that traumatic event at the
> speed of light.

> Therefore, our chances of coming across such radiation
> – is probably very close to zero!

The cosmic background radiation, which was discovered by
chance, is a first-class scientific discovery. However, perhaps we
should look for its sources not necessarily in "the Big Bang," but
elsewhere.

If there was "a Big Bang," the radiation should have arrived
in a clear, significant way from a certain direction, rather than
homogenously and continuously from all directions.

Perhaps those "microwaves" tell a completely different
story:

> Perhaps that weak radiation which reaches us from all
> around is some evidence of the **flux of fracticles** com-
> ing to us from the universe around us, with more or less
> equal intensity from all directions.

> Perhaps the background radiation that has already been
> discovered, like radiation in different wavelengths that
> will still be discovered in the future, tell part of "the story
> of fracticles."

Perhaps all kinds of radiation, just like anything else in the universe, are a result of the "vortexes" created by the fracticles that come in a massive flux from all over.

The cosmic background radiation could offer some evidence that could support the theory of fracticles, but it would probably be difficult to attribute it to the "Big Bang."

Finite, or Infinite?

Moreover, the rate of expansion of the universe would automatically become very close to the critical rate determined by the energy density of the universe. This could then explain why the rate of expansion is still so close to the critical rate, without having to assume that the initial rate of expansion of the universe was very carefully chosen.

Stephen Hawking

The view of "gravitation" as "repulsion" through an **external** factor, rather than an "attraction" that derives from the bodies **themselves**, has several consequences, which may redirect our thought to a new, unusual direction, regarding the universe in which we live.

If we assume that we are in a "finite universe," where there are "countless" stars, but not an "infinite number" of stars, where there is "universal attraction," according to Newton:

> ➤ In this universe, the distances between the stars and galaxies will decrease with time[29] , each mass attracts each other mass, and it must finally **shrink and collapse**, after all the energy in it from the expansion will subside (according to the Big Bang Theory).

> ➤ If a "finite universe" had "universal repulsion," where every mass repulses every other mass, the universe would

---

29 Although, at this point in time, and in our universe today, distances are still growing further (editors).

have to **expand with accelerated speed**, where every galaxy would distance themselves from their neighbors at an increasing speed. This conclusion is probably being reinforced by observations of astrophysical research conducted in recent years.

> Since these observations and research that pertain to the "redshift" of distant galactic light, point to the idea that we live in a universe that is expanding at an accelerated rate[30], rather than a shrinking or static universe, we should instead prefer **"universal repulsion"** to **"universal attraction"** or to any other combination of **attraction** and **repulsion**.

> On the surface of every star or cluster of stars, an "attraction force" must apparently appear. This is **the repulsion** caused by the fracticles coming from all other stars and masses in the universe around them.

> In an "infinite universe," where there are an infinite number of stars and galaxies, the distance between the stars and galaxies would have remained without any significant change, according to both the "attraction theory" and the "repulsion theory," due to the balance of forces (attraction or repulsion) exerted on the masses from all directions. Each one of the stars in space would "see" equal forces of **attraction** or **repulsion** from all directions, and could not move significantly without violating this balance of forces.

If we indeed live in a universe that is expanding at an accelerated rate, we may draw a few more conclusions:

> The universe in which we live is "a finite universe"

---

30   For further reading, we suggest the work of **Alice Richard** and **Avishai Dekel**. (Appendixes: **"List of Notable People,"** footnote on Chapter 8: **"Background Radiation and the Big Bang"** (The editors).

("accelerated") rather than an "infinite universe" ("static").[31]

> In a "finite universe" such as ours, there is "universal repulsion" **and not** "universal attraction."

> The repulsion in the universe may have derived from "a flux of fracticles" coming towards each one of the astronomical objects from all other astronomical objects and from the fracticles that are already in intergalactic and interstellar space. This is because as the "flux of fracticles" hits every one of the astronomical objects, it "tries to push it" and distance it from its neighbors.

> The "pressure" created by the "mutual pushing" of each one of the astronomical objects causes the universe to expand to increasing dimensions at an accelerated rate.

> Perhaps we may renounce the Big Bang Theory as the only explanation for the creation of the world, and maybe we should even change it or replace it.

"The End of the Universe"

> The end of the universe will probably not happen as a result of the Big Crunch, but rather when these astronomical objects distance themselves from each other and reach the speed of the fracticle flux.

> A flux that does not collide with anything will disappear in an infinite empty space.

---

31 As is customary today, researchers who examine the structure of the universe tend to think of it as existing according to three main possible models: **"A Closed Universe,"** which will eventually shrink towards "The Great Collapse," since the density of matter in it is above the critical density; **"A Flat Universe," which continues to expand but at a** decelerated rate, because of "gravity that is too small," as the quantity of matter in it is lower than the critical density; and an "Open Universe" that continues expanding forever at an accelerated speed, a model that has several properties that resemble **"The Finite Universe,"** proposed in **"Attraction – Secrets of Gravity"** (editors).

> It is likely that (in the distant future) the expansion rate of the universe will reach the speed of fracticles, which means, **the speed of light**.

> In such a case, there will no longer be any relations or connections between the various astronomical objects, and the universe, as we currently think of it, will cease to exist.

> If there will be no external "fracticle pressure" on the astronomical objects, they may explode, due to the "fracticle pressure" created by the "sub-atomic" fracticles or by the vortexes of which they are composed of.

> This "explosion," which could occur simultaneously or within a short period of time all over the expended universe, could replace the accepted model of "the Big Bang."

> Perhaps every "exploding" astronomical object would be the beginning of a new universe, which is not connected in any possible way to the other universes that were just "born" or "created."

If intellige
nt life were to appear on the surface of one of the planets in the "newborn" universes, then "their" universe would be, in their eyes, "**the only universe**," without being able to observe any other universe.

CRITIQUE

In his lectures: *The Feynman Lectures on Physics*[32], Richard Feynman describes gravitational theories that are similar to the fracticle theory, where fast particles reaching Earth from all directions create the effect of gravity in a fairly similar format to the one described in **Attraction - Secrets of Gravity**.

I add here the part of "Feynman Lectures on Physics Vol. 1 Ch. 7, The Theory of Gravitation" dealing with theories resembling the fracticle theory.

"7–7 What is gravity?

"*But is this such a simple law? What about the machinery of it? All we have done is to describe how the earth moves around the sun, but we have not said what makes it go. Newton made no hypotheses about this; he was satisfied to find what it did without getting into the machinery of it. No one has since given any machinery. It is characteristic of the physical laws that they have this abstract character. The law of conservation of energy is a theorem concerning quantities that have to be calculated and added together, with no mention of the machinery, and likewise the great laws of mechanics are quantitative mathematical laws for which no machinery is available. Why can we use mathematics to describe nature without a mechanism behind it? No one knows. We have to keep going because we find out more that way.*

---

32   Feynman, Richard P. *The Feynman Lectures on Physics, Vol.1*. Addison Wesley, 1963.

"*Many mechanisms for gravitation have been suggested. It is interesting to consider one of these, which many people have thought of from time to time. At first, one is quite excited and happy when he 'discovers' it, but he soon finds that it is not correct. It was first discovered about 1750. Suppose there were many particles moving in space at a very high speed in all directions and being only slightly absorbed in going through matter.*

"*When they are absorbed, they give an impulse to the earth. However, since there are as many going one way as another, the impulses all balance. But when the sun is nearby, the particles coming toward the earth through the sun are partially absorbed, so fewer of them are coming from the sun than are coming from the other side. Therefore, the earth feels a net impulse toward the sun and it does not take one long to see that it is inversely as the square of the distance—because of the variation of the solid angle that the sun subtends as we vary the distance. What is wrong with that machinery? It involves some new consequences which* **are not true.** *This particular idea has the following trouble: the earth, in moving around the sun, would impinge on more particles which are coming from its forward side than from its hind side (when you run in the rain, the rain in your face is stronger than that on the back of your head!). Therefore there would be more impulse given the earth from the front, and the earth would feel a resistance to motion and would be slowing up in its orbit. One can calculate how long it would take for the earth to stop as a result of this resistance, and it would not take long enough for the earth to still be in its orbit, so this mechanism does not work. No machinery has ever been invented that 'explains' gravity without also predicting some other phenomenon that does not exist.*"

According to Feynman, these theories are not true, since as Earth revolves in orbit around the sun, it collides with these particles, and the collisions with them in the direction of its motion are more frequent and faster than the collisions with particles coming from the other directions.

This "particle resistance" to the motion of Earth was meant to slow down its rotational speed around the sun, so that in the end, Earth would fall into the sun.

Since, as we all know, Earth continues to rotate in orbit around the sun, these theories, **according to Feynman**, are incorrect.

The critique

What is the **difference** between the theories described by Feynman and the **Fracticle Theory** that appears in *Attraction - Secrets of Gravity*?

The difference between the two kinds of theories is substantial, and results from the fracticles themselves and their speed of motion.

According to the "Feynman Theory," there are some particles, with undefined mass, which move at some fast speed, which is apparently, less than the speed of light.

"Fracticles," by comparison, are several orders of magnitude smaller than any known particle, and probably have no mass or almost no mass (as mentioned earlier). Fracticles create mass through vortexes or other clusters of fracticles that disturb the free passage of the flux of other fracticles that are composed of singular units, which move in straight lines at the speed of light – the flux – until they collide with a cluster, such as the fracticle vortexes.

Light and electromagnetic waves are among the phenomena "created" by the fracticles, and this is described as "a cyclical disturbance to the flux of fracticles" that propagates at the speed of light.

After the experiment conducted by Michelson and Morley, it was revealed that objects that come across electromagnetic radiation (the visible light is just a narrow part of it) will encounter it **always at one, singular speed**, the **speed of light**. This is even if they are moving towards the source of radiation, perpendicular to it or away from it.

When the Earth moves around the sun at a speed of about 30 km per second, it will still be hit by all the "fracticles" at the speed of light, regardless of the direction from which they are coming.

If the flux of fracticles is more or less equal from all directions, the number of collisions Earth experiences each second will be equal in all directions.

If the size of fracticles is more or less equal and their speed is constant, the momentum of every collision of a singular fracticle with a vortex, would equal the momentum of the collision between another fracticle coming from the opposite direction.

If the general momentum experienced by Earth is equal in all directions, there is no reason why it should decelerate its movement around the sun, as described by Feynman when he refers to large particles that move at speeds lower than the speed of light.

Please note, Earth's movement around the sun is also caused by "fracticles," and thus it is unlikely that they are both the ones to cause its motion and to stall it at the same time.

Will all fracticles hit Earth with the same energy?

"Electromagnetic radiation" from stars and galaxies, will

hit Earth with different energies and their momentum will also be different. This is due to the "Doppler effect," which causes a "blueshift" in the direction of Earth's motion or a "redshift" in the opposite direction.

Electromagnetic radiation reaches Earth from all stars, nebulas, galaxies, radio sources, background radiation in microwaves and radiations at other wavelengths, from bodies that have already been discovered and from those that have not yet been discovered. These radiations may indeed slightly slow down the Earth's rotational speed around the sun.

Fortunately, Earth has "advantageous conditions," since the sun loses close to five million tons of its mass per second by radiation (only a minuscule part of which reaches us).

The loss of mass involves a weakening of the sun's gravity (the Shadow Cone), which reduces the risk that the Earth will fall into the sun due to the electromagnetic radiation of all bodies in the universe.

The heavy planets, which are in orbits further away from the sun than the orbit of Earth, also affect our distancing from the sun.

Such "advantageous conditions" may not "help" the moons of other planets, thus the orbit may get smaller, though at a very slow rate, according to the relation between the electromagnetic radiation and the great, undisturbed flux of fracticles that is responsible for gravity.

So, we can see that the fracticles theory is still alive and kicking…

Epilogue, or the
Beginning of a New Road

"If we were not able or did not desire to look in any new direction, if we did not have a doubt or recognize ignorance, we would not get any new ideas."

Richard Feynman

"There is no doubt that every forward movement will give rise to questions that will need to be answered, and new riddles will appear that we will need to solve. We remember Pascal when he said that the equation of man will continue to increase like a snowball with no end. The larger it becomes, the more points there will be to examine abut the unknown."

Shmuel Sambursky

As the book comes to an end, I would like to share some of my thoughts, doubts and reasons that motivated me to write it. These may help the readers reveal their own overt or covert motivations, or points in which our ideas are similar or close, so that we may together succeed in expanding our joint framework of thought and develop additional channels for new ideas.

Even if we do not get to "reveal the truth," the existence of which is doubtful, we may be able to bring about useful advancements in the fields that we have discussed and the fields that seemingly are not directly part of the themes of this book.

I would like to ask the readers to keep thinking, talking, discussing and pushing forward the ideas they embraced while reading the book, or which arose while reading the book, and so to help to widen the openings that may have appeared.

Quite a few of the subjects I wrote about, or conclusions I have pointed to, could possibly serve as starting points for scientific work or research that would reach much further than I have.

The readers are welcome to use the ideas in the book, if they find them useful to their work, and I would be happy to thank them.

More than fifty years have passed since I started thinking about the essence of "gravity" and the option of giving it a reasonable explanation, beyond the definition of "attraction force" that derives from the "attracting" bodies themselves or from mass creating "curved space-time."

I could not find any logical way to explain the **"attraction mechanism"** or gravity based on straight movements of any material bodies, and finally I had to completely abandon the idea that attraction derives from **the bodies themselves**.

Since I could not find any possible mechanism to explain how "attraction force" works, I started wondering if we could be mistaken in the interpretation we give to phenomena whose sources we do not know, which is why we explain the phenomena by an attraction force that derives from the bodies themselves.

Perhaps the source of the observed phenomena is different.

I was also troubled by questions about the permanent and borderline speed of light, the essence of time and the riddle of motion.

It seemed strange to me that a certain force can make bodies move towards an electric wire, perpendicular to the direction of the current running through it, which is why I presented a proposal for a different theory that explains this phenomenon through the "tree frogs tale." It was also hard for me to accept the existence of "curved force lines" within a magnetic field.

The concept of "field," if it is devoid of any "thing" that moves, seemed to me to be vague, and even somewhat mystical, as did concepts such as "force" or "energy," which were hard to see as belonging directly to reality.

Just like others, I was bothered by the question:

Could it be that there is a simple connection between gravity, magnetism, electromagnetic forces and the forces within the atomic nucleus?

Could it be that such a connection escapes us because nature is "playing hide and seek" with us, or because accepted theories are the ones hiding it from us?

Could it be that the universe around us "willingly" provides us with answers to the questions that we present, as long as we are open enough to a new way of thinking, and do not rely on old, accepted answers?

Could "mass," "energy," "force" and "speed" be part of a real reality, or are they intellectual models that we have developed and turned into accepted "facts"?

Does the universe have an end in space? Does time have an independent physical entity?

What is motion, without which we would not be able to see what happens around us, and without which, bodies would have no volume or any other property, and so, our existence and the existence of the entire universe depends on it more than on anything else?

What is the relation between motion and memory, or what is the connection between physical phenomena and the way we think of it?

Would answers to these questions open a new channel to understanding concepts that the current outlook conceals or blurs?

During the writing of this book, I delved deeper and deeper into these questions, trying, as best as I could, to offer answers that match my way of thinking, or to present them in such a way that would lead the reader to take part with me in the search for new answers to these questions.

As writing progressed, especially in the chapter discussing the limitations of human thought, I started to feel that perhaps I was not dealing with physical problems from reality, but with our way of thinking, about ourselves, and our surrounding.

Do we have any direct connection to reality, or do our senses mislead us?

Is the "software" running in our minds not the biggest obstacle that comes between us and reality?

Is it possible to overstep, even just slightly, from the limitations of our way of thinking?

I started wondering whether we had, unintentionally, become caught in a certain interpretation of our surroundings, with an interpretation that had ceased to serve us with the desired efficacy and due to which we now walk on a long, winding side lane, instead of looking for a wide, straight main road.

In this book, I tried to build a uniform theory, in which "a flux of fracticles," which comes from all bodies around us or which was created in the space between them, serves as the basis for all the complexity and beauty of the world of phenomena that we experience through our senses.

I believe that I have managed to show that the quality explanation offered by the Theory of the Flux of Fracticles for gravity and its connection to all other phenomena is preferable over existing explanations.

I think it is important to emphasize that the flux of fracticles is different from the ether theory that was popular until the beginning of the twentieth century, as fracticles move at the speed of light, in relation to any known material body.

This flux of fracticles replaces the concept of "force" in all of its forms, and provides our perception of the world with a very tangible basis, which, to my opinion, is simpler and a better basis than the accepted perceptions.

The fracticles that compose the flux **have no properties**, since the attribution of a certain property to an elementary particle **would necessarily remove it** from the group of **elementary** particles.

Every property that we attribute to a certain particle has a source or a reason that makes it appear and thus the causes of the property become its basic building blocks. Therefore, **a particle that has properties cannot be considered an elementary particle.**

The multiplicity of "elementary" particles that constitute the "standard model" of the atom, as it is accepted today, may indicate more than anything, that we are not on the main path to revealing the secrets of matter and the universe, and perhaps we should change the way we think, to a greater or to a lesser extent.

I have endeavored to show that one can find alternative ways to explain the structure and formation of the atom, though I am not sure that these are the best ways.

I have tried to redirect the readers' "attention" into a different direction than the accepted one. This is so they would share that feeling of discomfort regarding the direction in which the research of the "families" of the elementary particles of matter has become that of "families with too many children."

Are we indeed researching elementary particles?

In the chapter that discusses the speed of light, I tried to show that the speed of light does not change when it passes through a vacuum, air or any other transparent medium. The only thing that changes is the direction or the length of the path that the light passes.

I reached the conclusion that perhaps the speed of light is the only speed in our world, but is this the only possible conclusion?

The repulsion theory that I presented, leads to a possible conclusion that the "borders" of our universe are finite, but they continue to expand at an increasing speed, and in the far future they will approach the speed of light, where the distinction between the finite and infinite loses all meaning.

The extent of my success in finding a new or different way of thinking depends, more than anything else, on the readers themselves, on their openness and their willingness to pass with me through the obstacles I have had to overcome, through the ones I tried to overcome and through the ones I failed to pass.

During the writing of this book, I tried to indicate at the solution that seemed, to me, to be the most suitable for some of the questions I dealt with.

In other questions, I let the readers "struggle" on their own with the problems, in order to turn them into partners in the search for a new way. With regard to some questions, I did not want to set a direction that might not offer a simpler answer than the accepted theories.

If my way of thinking, with the questions that I raised, matches your way of thinking to some extent, I would be pleased to see you as a full partner on the path where **doubt** illuminates possibilities that were pushed aside, due to the blinding light of "accepted truths."

If I caused any of my readers embarrassment by raising questions or answers that do not match their way of thinking or the accepted way of thinking, I do apologize.

I hope that *Attraction – Secrets of Gravity* will contribute somewhat to the development of thought in the next few decades.

There is no truth in this book.

It is a search for simple ways to understand the world around us, and an attempt to understand how we look for these new ways.

The more new ways we discover, the greater our chances of finding answers that may serve us better – until other, better ways replace them.

Will we be brave enough to walk along at least one of these unknown paths?

Reuven Nir
Kibbutz Mechavia, Israel - 1998

Appendices

- ➤ Impressions
- ➤ Recommended Bibliography
- ➤ List of Notable People
- ➤ Glossary
- ➤ Sources of Citations & References

IMPRESSIONS

Orit and Herzl Bodinger

The pressures of life have kept us from passing on our impressions directly after reading your wonderful manuscript. Although we received it at a very difficult period, we both read it and could not put it down, just like reading a thriller.

It would be pretentious and silly of us to give a sophisticated or scientific opinion of the physical theory you develop in *Attraction – Secrets of Gravity*.

We still have not seen any "tree frogs" nor a "flux of fracticles" in our own eyes, so it would be hard for us to testify to their existence, and we will leave that to greater people than us.

We would like to recommend the amazing reading experience you have given us, as we enjoyed the flowing and charming style of your writing, your wonderful illustrations and the original expressions, some of which have already become folklore in our home.

The book has made us very curious, eager to look at everything we know through a completely different perspective. In our eyes, *Attraction - Secrets of Gravity* is just as exciting as Stephen Hawking's *A Brief History of Time*.

The ability to think differently, to examine what we take for granted and to challenge it – is a huge emotional and intellectual celebration.

Even if the book remains just in the field of philosophy, it is epic, as it invites the reader to have a "spiritual revolution" in his outlook of this world.

We wish you much luck, just like in Nathan Zach's poem "How Does One Star Dare."

Orit and Herzl Bodinger

Ramat Hasharon, Spring 1999

Orit Boger Bodinger

Actress, and a playwright.

Graduate, of the Theater Department of Tel Aviv University.

Founder of the community educational theater: "Beruach Hazman" ("In the Spirit of the Times"). Wrote the plays: "The Motorcycle" and "Close."

Herzl Bodinger

Commander in Chief of the Israeli Air Force: during the years 1991-1996. Commanded the First Fighter Squadron, the Ramat David airbase and served in senior staff positions.

Graduated from, Bar-Ilan University, with majors in Economics and Business Administration.

Currently: President of RADA Electronic Industries.

Shulamith Kreitler

The more I progressed with the reading of the manuscript, the more I was "drawn" into it.

It is a rare occasion when a reader has the opportunity to read a truly creative book. It is a rare event due to the limitations of human creativeness in general and due to social restrictions that limit the expression of creativity and its appearance in reality.

Even if someone has a significant creative potential, there is no certainty that he will be discovered. There are other traits that are required, such as daring, persistence, education and expertise on the one hand, and the willingness of society to encourage and accept the expressions of creativeness on the other hand. Therefore, we must celebrate the publication of this book.

As one reads **Attraction – Secrets of Gravity**, one is confronted with bare creativity. The author, without hesitation or avoidance, simply raises fundamental questions for discussion regarding "the essence of reality," "time and space," "mass," "force," "speed," "the birth and death of the universe," "limitations in recognition and in human thought," among others.

As questions were raised and discussed, new and old conventions were destroyed with almost "unbearable ease," with elegance and convenience that derive from the innovative perspective offered in the book.

And yet, the book is readable and at times even amusing. The most incredible ideas are presented in simple, direct language, often as quotes, metaphors, anecdotes or fables that help to demonstrate ideas, beyond their own depth and poetical beauty.

However, it is no coincidence. It is exactly the closeness between the simple language and the ease of skipping from daily life to the physical context, and the intimate connection between the scientific statement and its philosophical implications, which are the main signs of creativity.

Thus, **Attraction – Secrets of Gravity** becomes not only intellectually interesting, but an important book.

Such books serve as the fertile ground from which new scientific developments may emerge.

Whether that happens or not, let us make a wish:

May we all continue to write, publish and read such rare and wonderful book.

Shulamith Kreitler
Tel Aviv, Spring 1999

Shulamith Kreitler

Professor of psychology at: Tel Aviv University, researcher, lecturer and therapist. Some of her books include *Psychology of the Arts, Cognitive Orientation and Behavior* (with her husband, Prof. Hans Kreitler) and *The Psychology of Symbols*. She has published more than a hundred scientific papers. She currently heads the Unit of Psych-oncology at Tel-Aviv Sourasky Medical Center, which combines advanced technology, research, innovative psychological techniques and using mental strength in the fight against cancer and in curing cancer patients.

RECOMMENDED
BIBLIOGRAPHY[33]

Agassi, Yosef / *The Ongoing Revolution. The History of Physics from Ancient Greece until Einstein 1968*, Dvir Publishing House, 1976.

Articles / *Scientific American*, special issue. Volume 3, November 1991, p. 37, 63.

Bekenstein, Jacob / *Black Holes and the Limits to Information*, Galileo, Scientific Magazine, Volume 35, July – August 1999.

Ben Noon, Chemi / *The Limits of Reason - On Thought, Science, and Faith - Conversation with Isaiah Leibovich and Joseph Agasi*, Keter, 1997.

Bergson, Henri-Louis / *Time and Free Will* (*Essai sur les données immédiates de la conscience*), 1889.

Calder, Nigel / *Einstein's Universe*, 1996.

Descartes, René / *Meditations on First Philosophy, 1641*.

Einstein, Albert / *The Foundation of the General Theory of Relativity 1916*.

Einstein, Albert; Infeld, Leopold / *The Evolution of Physics: The Growth of Ideas from the Early Concepts to Relativity and Quanta*. Published by, Cambridge at the University Press, 1938.

Feynman, Richard / *The Meaning of It All: Thoughts of a Citizen Scientist*, 1963.

---

33   This is not a list of references for **Attraction- Secrets of Gravity**. It is a recommendation by the editors and the author for readers who are not acquainted with the topics of discussion to expand their knowledge. It is not a comprehensive list of all relevant bibliography.

Fraser, Gordon; Lillestol, Egil; Sellevag, Inge / *The Search for Infinity: Solving the Mysteries of the Universe*, 1995.

Gamow, George / *The Birth and Death of the Sun*, 1940 (revised 1952).

Gleick, James / *Chaos: Making a New Science*, 1987.

Hawking, Stephen / *A Brief History of Time: From the Big Bang to Black Holes*, 1988.

Hawking, Stephen / *Black Holes and Baby Universes and Other Essays*, 1993.

Internet Website / Einstein's Dreams - http://www. einsteindreams.com

Jeans, James Hopwood / *The Stars in Their Courses, 1931.*

Kant, Immanuel / *The Critique of Pure Reason*, 1781.

Kramer, Daniel / *The Expansion of the Universe, the Universe Presses the Accelerator.* Galileo, Scientific Magazine, Volume 35, July – August 1999.

Leibowitz, Yeshayahu / *Between Science and Philosophy.*

Leibowitz, Yeshayahu / *Body and Mind: The Psycho-Physical Problem*, 1990.

Newton, Isaac / *Philosophiæ Naturalis Principia Mathematica* (From Latin: «Mathematical Principles of Natural Philosophy»), 1687. Berkeley: 1947.

Poincaré, Jules Henri / *Science and Hypothesis*, 1952.

Restak, Richard M. / *Brain: The Last Frontier: An Exploration of the Human Mind and Our Future*, 1979.

Sambursky, Shmuel / The Evolution of *Physical Thought from the Presocratics to the Quantum Physicists: An Anthology*, 1972.

Steiner, Rudolf / *Human and Cosmic Thought*, 1914.

Thorn, Alice / *The Story of Madame Curie*

List of Notable People[34]

ALPHER, RALPH ASHER (1921 – 2007)

An American cosmologist, He worked with George Gamow on the Big Bang Theory and edited all the calculations, including the temperature of the background radiation.

ANAXAGORAS (APPROX. 500 - 425 BC)

A Greek philosopher, who taught in Athens. His substance theory is based on the idea that "Each thing contains in itself parts of other things." He said that "there is a portion of everything, i.e., of every elemental stuff, in everything," but "each is and was most manifestly those things of which there is most in it."

ARCHIMEDES (287 - 212 BC)

A Greek mathematician and scientist, one of the greatest mathematicians of ancient times ("Give me a place to stand on, and I will move the Earth"). One of his essays is the source of our knowledge of the heliocentric world, of his contemporary, Aristarchus of Samos.

ARISTOTLE (384 – 322 BC)

A Greek philosopher, scientist, researcher and biologist. Plato's student, and the teacher of Alexander III

---

34 The list is in alphabetical order. People who are still alive were mentioned without their dates of birth.
Sources: *Physical Thought from the Presocratics to the Quantum Physicists: An Anthology, A Brief History of Time, The Encyclopaedia Hebraica, Galileo* (scientific magazine), Magnes Publishing House, *The Limits of Reason - on Thought, Science, and Faith-Conversation with Leibowitz, Yeshayahu and Joseph Agasi, National Geographic no. 17*, search engines, websites, libraries and online encyclopedias, the author and the editors.

of Macedon ("Alexander the Great"). He founded his own school, the Peripatetic school. His writings established the teleological (purposeful) approach in natural sciences and in ethics, and had a great influence on both scientific thought and on ethical approaches up to the new era.

BEKENSTEIN, JACOB (1947 – 2015)

An Israeli professor of physics. He became the head of the Hebrew University's theoretical physics department. He was elected to the Israel Academy of Sciences and Humanities and the International Gravity Field Service (IGFS).

His doctoral adviser was John Wheeler and he worked with Stephen Hawking. He was awarded the Rothschild Prize, the Landau Prize, the Ernst David Bergmann Prize, and the Gravity Research Foundation in the US. He was the first to prove "black-hole thermodynamics" and was one of the formulators of the "Generalized Second Law of Thermodynamics." He assumed that a black hole emits radiation, has a temperature, decreases in mass and grows in entropy.

BERGSON, HENRI-LOUIS (1859 – 1941)

A French philosopher who placed intuition at the center of existence.

Considered an "irrationalist" by many. He addressed and researched concepts such as "time," "duration," "continuity," "simultaneity," "quality," "quantity" and more. He tried to show that the source of problems and contradictions in human thought is the confusion in the interpretation of these concepts. His books include *Time and Free Will: An Essay on the Immediate Data of Consciousness, Matter and Memory, Creative Evolution* and more. He awarded the Nobel Prize in Literature.

BOHR, NIELS HENRIK DAVID (1885 – 1962)

A Danish physicist, one of the greatest theoreticians of atomic, and nuclear physics. His work, *On the Structure of Atoms and Molecules*, constitutes an elementary contribution to the combination of atomic theory and quantum theory. He published a series of important papers on the philosophical significance of quantum mechanics, and introduced the concept of the complementarity principle (from complementary) as a generality of the concept of causality. In 1922, he was awarded the Nobel Prize in physics. (See Heisenberg, Werner, and Planck, Max.)

BOLTZMANN, LUDWIG EDUARD (1844 – 1906)

An Austrian physicist, and philosopher. He developed statistical mechanics and clarified the relations between entropy and probability of a thermodynamic state of a system. His book, *Lectures on Gas Theory,* was published in 1896.

BRUNO, GIORDANO (1548 – 1600)

An Italian philosopher, and a Dominican monk. He was one of the avid supporters of the Copernican model, and promoted it in his travels throughout Europe. He was burned at the stake by the Inquisition. In his book *On the Infinite, Universe and Worlds*, he raised the idea that the fixed stars that fill infinite space are suns like our sun. (See Copernicus, Nicolaus and Galilei, Galileo).

CAVENDISH, (SIR) HENRY (1731 – 1810)

An English physicist, chemist and misanthrope, who was born in France. He discovered hydrogen in 1766. He became famous in 1798, when he conducted the first accurate experiment to measure the force of gravity between masses in the laboratory and the first experiment to yield accurate values for the "gravitational constant." He conducted the experiment using the torsion balance apparatus, which he improved.

COPERNICUS, NICOLAUS (1473 – 1543)

A Polish astronomer and scholar, founder of the heliocentric theory, which paved the way for a new era in science. His book, *On the Revolutions of the Heavenly Spheres*, which he worked on for 33 years, was published shortly before his death. (See Ptolemy, Claudius).

COTES, ROGER (1682 – 1716)

An English mathematician. He worked with Newton on the second edition of the Mathematical Principles of Natural Philosophy, and wrote its introduction. His interpretation of Newton's theory and his opinions were typical of the dogmatic picture the "Newtonian school" gave its founder's views.

CURIE, MARIE (MARIA SKŁODOWSKA) "MADAME CURIE" (1875 – 1934)

A chemist and physicist who was born in Poland, immigrated to France and married the physicist Pierre Curie. She and her husband were groundbreaking researchers of radioactive research of uranium and thorium, and discovered the elements polonium and radium. In 1903, she was awarded the Nobel Prize in Physics, together with her husband and Prof. Henri Becquerel. In 1911 she was awarded the Nobel Prize in Chemistry. She headed the Red Cross Radium Institute, and she was a professor of chemistry at the Sorbonne. She died of aplastic anemia, probably due to overexposure to radiation.

DA VINCI, LEONARDO (1452 – 1519)

An Italian polymath. The most prominent person in the Renaissance. He had an ability to create and a critical approach that are admirable throughout the generations. He was a painter, sculptor, architect, inventor, planner and strategist, and invented flying and war machines hundreds of years before their time.

Da Vinci was an eager and thorough researcher of nature, who laid strict foundations for the sciences that blossomed after his time. He preceded Copernicus, Newton and many others with his *"heliocentric"* view of our solar system, "gravity" and other ideas. Among his famous paintings are the Mona Lisa, the Last Supper mural, The *Virgin* of the *Rocks,* The Virgin and Child with Saint Anne and many outstanding drawings, decorated with his explanations in his famous "mirror writing." (See Michelangelo Buonarroti).

DESCARTES, RENÉ (1596 – 1650)

A philosopher, and mathematician who was born in France. Many consider him to be the founder of the New Philosophy. One of his greatest discoveries was analytic geometry. He wrote many books in mathematics, physics and philosophy. The most famous of these include *Discourse on the Method, Meditationes de Prima Philosophia (Meditations on First Philosophy)* and *Principia Philosophiæ (Principles of Philosophy)*, among others.

DEKEL, AVISHAI

An Israeli professor of astrophysics at the Hebrew University. Led a group together with Dr. Idit Zehavi on a parallel research to the research conducted by the Richard Ellis group on similar phenomena (*Nature* 1999, *Haaretz*, June, September 1999). His group collaborated with Ellis's group, and results from the research conducted by both groups show that the universe is expanding at a rapid pace. An assumption also arises regarding the existence of an intergalactic "repulsion force." This is similar to Einstein's conclusion ("the cosmological constant," "antigravity," "Lambda"), which he would later dismiss when speaking to Gamow as "the biggest blunder of my life." Today, new assumptions try to explain the accelerated expansion of the universe. For instance, that empty "space-time" has "its own repulsion force" that differs from "gravity" both in its direction and by its nature, and also that there are two forces of gravitation in the universe: repulsion and attraction. (See: Ellis, Richard).

DEMOCRITUS, (460 – 370 BC)

A Greek philosopher, developed the atomist concepts based on Leucippus.

EINSTEIN, ALBERT (1879 – 1955)

A Jewish German, and one of the greatest physicists in the modern era. He founded *Special Relativity* in 1905 when he was 26 years old. He adopted the hypotheses of Riemannian geometry in the development of *General Relativity* in 1915 at the age of 35, which formulated a new "Theory of Gravity." His work on "The Theory of Light Quanta" is one of the most significant contributions to quantum theory, after Plank's work. His theoretical research on Brownian motion is irrefutable proof for the molecular structure of matter. He formulated the relations between energy mass and the speed of light in his famous formula: $E = M C^2$. Received the Nobel Prize in Physics in 1921.

ELLIS, RICHARD SALISBURY

A British, Astrophysicist from Cambridge University. Together with scientists from the University of California, Berkeley, he led research on the accelerated distancing of galaxies, by observing "supernovae" (*The Astrophysical Journal*, June 1999).

He collaborated with Dekel Avishai's group from the Hebrew University in this field. (See Dekel, Avishai).

EPICURUS (341-271 BC)

A Greek philosopher, Founder of the school of philosophy called Epicureanism, which developed the atomic doctrine, following in the steps of Democritus.

ERHARD, WERNER

An American philosopher. Founder of the "Est" and "Forum" workshop techniques. His approach is practical, dynamic and based on the integration of philosophies such as "Zen Buddhism," "Existentialism," "Logo-therapy," "The Dale Carnegie Technique," "Chazal Torah" and more, up to programs that enable sustainable changes with daring and creativity in one's relationships and quality of life.

EUCLID OF ALEXANDRIA (FOURTH CENTURY BC)

A Greek Mathematician, worked in Alexandria during Ptolemy I Soter. Considered the greatest of the Greek mathematicians. He developed "Classic Geometry" in his essay *The Elements*. Euclidean geometry is based on "intellectual axioms" (especially "parallel postulate"), from which the rest of its laws are logically derived. It is especially suited for use in small scale systems. In modern scientific research, Riemann and others replaced these axioms and got a different kind of geometry, where parallel lines can meet in infinity (the hyperbola model) or where there are no parallel lines in reality, since a straight line closes itself as a never ending circle (the elliptic model). His books discuss division, optics phenomena, and mistakes.

FARADAY, MICHAEL (1791 – 1867)

A British physicist and chemist, one of the greatest experimental scientists, founder of "The Theory of the Physical Field," which abolished the concept of action-at-distance. Some of his greatest discoveries include "electromagnetic induction," the elementary "laws of electrochemistry" and the "rotation of a plane of polarized light" by the "electromagnetic field." In 1851, he summarized his ideas regarding "the reality of force lines." (See Maxwell, James Clerk).

FEYNMAN, RICHARD PHILLIPS (1918 – 1988)

An American physicist, took part in the Manhattan Project, was awarded the Nobel Prize together with his colleagues Sin-Itiro Tomonaga and Julian Schwinger, for their contributions to the development of quantum electrodynamics. He was a pioneer in his theoretical work on the super-fluidity **of super**-cooled liquid helium, beta decay **theory, collision processes at high energy levels and the development of** the quark theory. He developed techniques for calculating and registering physicality, including the Feynman formulae. He was a senior investigator of the Challenger disaster. He was a teacher, an artist and a public figure. His books include *The Feynman Lectures on Physics, The Character of Physical Law,* and *QED: The Strange Theory of Light and Matter.*

GALILEI, GALILEO (1564 – 1642)

An Italian scientist, and astronomer. He championed Copernican theory and established it through his findings with the telescope. His book, *Dialogue Concerning the Two Chief World Systems,* which was banned by the church, led to him famously being trialed by the Inquisition. His research about the pendulum, the law of free fall and other mechanical problems were published in his book *Dialogues Concerning Two New Sciences,* and helped establish modern mechanics. (See Copernicus, Nicolaus, Bruno, Giordano).

GEORGIY ANTONOVICH GAMOV (GAMOW, GEORGE) (1904 – 1968)

Born in Odessa and gained his education as a physicist, mathematician and astronomer in Russia. He was awarded many prizes, and published more than a hundred different scientific papers. He collaborated with Niels Bohr, Ernest Rutherford,

Edward Teller, Robert Oppenheimer and many others. He was a professor at many universities, including the University of Copenhagen, the University of Leningrad, Cambridge University, George Washington University and many others. Founder a theory of the internal structures of red giant stars and researcher of the structure of the living cell. A successful science writer, whose books include *Mr. Tompkins in Wonderland*, *Mr. Tompkins Explores the Atom*, and *The Birth and Death of the Sun*, among others.

In 1948, Gamow and his student Ralph Alpher produced an important paper on cosmogony entitled *The Origin of Chemical Elements*, which outlined how the present levels of hydrogen and helium in the universe could be largely explained by reactions that occurred during the Big Bang. They also forecast the background radiation.

GILBERT, WILLIAM (1544 – 1602)

A British scientist, doctor and researcher. He was the physician of the queen of England. He specialized in researching magnetism and electricity and laid the foundations for future research in these fields of physics. His book *On the Magnet and Magnetic Bodies* describes the magnet and the magnetism of Earth, based on systematic experiments. He influenced the work of Johannes Kepler, among others.

HAWKING, STEPHEN WILLIAM

A British mathematician, and physicist. One of the most prominent cosmologists today. His work includes developing assumptions on the properties of black holes, the birth and death of the universe, the structure of the universe, space and time, among others. He suffers from Lou Gehrig's disease, which started in his twenties, has made him completely paralyzed, and led to a loss of his ability to talk. He speaks with the aid of a speech synthesizer. His books include *A Brief History of Time, as well as, Black Holes and Baby Universes and Other Essays.*

As required by Cambridge University regulations, Hawking retired as Lucasian Professor of Mathematics in 2009, a position that was once filled by Newton.

HEISENBERG, WERNER KARL (1901 – 1980)

A German theoretical physicist. Founder of quantum mechanics, together with Max Born and Ernst Pascual Jordan. Formulated the uncertainty principle in atomistic phenomena, which also has an elementary meaning in scientific epistemology. He stated that our ability to precisely measure complementary sizes such as the position versus the momentum of the electron, or the energy of its state versus its life span, is limited. He remained in Germany during World War Two, but refused to take part in the development of the German atom bomb. He laid the foundations for the atom nucleus being composed of neutrons and protons. He discovered the allotropic forms of hydrogen. Awarded the Nobel Prize in Physics in 1932. (See: Bohr, Niels and Planck, Max).

HERTZ, HEINRICH RUDOLF (1857 – 1894)

A German physicist. His research of the spread of electric waves proved Maxwell's "electromagnetic theory of light." His work served as a starting point for the development of the wireless telegraph. The unit of frequency in the International System of Units is named after him.

HOYLE, (SIR) FRED (1915 – 2001)

An English cosmologist. At the end of World War Two, Hoyle, Gold and Bondi started developing a new theory about the universe, and in 1946 they made a breakthrough in the field and published their "Steady State Theory," which offered an alternative model to the Big Bang. Hoyle was the one to coin the term The Big Bang, as a pejorative to the idea that the universe started with a giant explosion.

HUBBLE, EDWIN POWELL (1889 – 1953)

An American astronomer. Started his career as a lawyer. His main work was conducted at the Palomar and Mount Wilson Observatories in California. He proved that the spiral nebulae are galaxies that are far beyond the boundaries of our Milky Way. Together with Milton L. Humason, he discovered the phenomenon of "Redshift" (according to the "Doppler effect" principle) of the light of galaxies beyond the Milky Way, and stated that they are "getting further away from each other in speeds that are proportional to their distances." This constant is named after him. The Hubble-Humason Law served as a basis for the theory on the expanding universe and for all the innovative cosmological perceptions. His books include *The Observational Approach to Cosmology*, and *The Realm of the Nebulae*

JEANS, (SIR) JAMES HOPWOOD (1877 – 1946)

An English mathematician, physicist, astronomer, researcher and writer. He taught at the universities of Cambridge and Princeton. He undertook research at the Mount Wilson Observatory in Pasadena, California. Specialized: in the research of radiation and thermodynamics. Fellow of the Royal Society in London. He criticized Laplace's theory on the creation of the solar system, and made an alternative assumption. Among his books are *The Dynamical Theory of Gases, Theoretical Mechanics, Mathematical Theory of Electricity and Magnetism, The Universe Around Us, Through Space and Time* and *Stars in Their Courses*.

KANT, IMMANUEL (1724 – 1804)

A German scientist, teacher and philosopher, who had an

immense impact on different fields in modern philosophy. He wrote about many subjects, including mathematics, physics, ethics, law, moral philosophy, geography, astronomy, geology and education. The 'father' of the theory of the creation of the solar system, as described in his book *Universal Natural History and Theory of Heaven*. The theory was named after him and Laplace. An author of: philosophical masterpiece essays. The central essay, *The Critique of Pure Reason*, described the centrality and freedom of reason in the creation of the external and internal reality, and its reciprocal relations with the ontological "being" that it will never get hold of. (See: Laplace, Pierre-Simon).

LAPLACE, PIERRE-SIMON (MARQUIS DE LAPLACE) (1749 – 1827)

A French astronomer, mathematician and physicist. Contributed to the development of the evolutionary change in the structure of the solar systemtheory. He developed his famous cosmogenic hypothesis in his book *The System of the World* (unaware of Kant's essay, forty years earlier). His book, *Analytic Theory of Probabilities,* is one of the classic essays in this field. (See Kant, Immanuel).

LEIBNIZ, GOTTFRIED WILHELM (1646 – 1716)

A German philosopher, mathematician, diplomat and lawyer. Invented, independently of Newton, differential and integral calculus. In contrast to Newton, he used concepts of relative space and time. He emphasized the importance of "the living force" in mechanics and the need for a teleological, divine principle to interpret physical phenomena. The starting point for his metaphysics was an endless multiplicity of non - material bodies that he termed "monads." His books include *Principes de la Nature et de la Grâce Fondés en raison, A New Method for Learning*

and Teaching Jurisprudence, New Essays on Human Understanding.
(See Newton, Sir Isaac).

LEIBOWITZ, YESHAYAHU (1904 – 1994)

A German-born Israeli microbiologist, researcher, publicist and philosopher. A professor at the Hebrew University. He specialized in the philosophy of science, religion and faith, values, history and more, and conducted theoretical and practical research, writing and editing. He also taught and gave lectures to the public. Leibowitz became famous thanks to his long-term work on questions regarding "body and soul." He dealt with the fundamental, existential questions of the State of Israel, the Jewish people and the Arab-Israeli conflict. His books include *Body and Mind, The Psycho-Physical Problem, Between Science and Philosophy,* and *Faith, History and Values.*

MACH, ERNST WALDFRIED JOSEF WENZEL (1838 – 1916)

An Austrian physicist and philosopher, one of the leading thinkers to introduce positivism in science.

His important book, *The Science of Mechanics,* influenced scientific thought at the end of the nineteenth century.

The Mach number was named after him, which is the ratio of the speed of a body to the speed of sound in the surrounding medium.

MAXWELL, JAMES CLERK (1831 – 1879)

A British physicist, founder of the mathematical theory regarding *"A Dynamical Theory of the Electromagnetic Field"* (Maxwell's equations) and the kinetic theory of gases.

He assumed that light is an electromagnetic wave, and he made optics part of electromagnetism. He showed that the expansion speed of the electromagnetic field equals the speed of light. He assumed that electric and electromagnetic forces operate within the "field." (See: Faraday, Michael).

MICHELANGELO DI LODOVICO BUONARROTI SIMONI (1475 – 1564)

An Italian, born in Florence. One of the world's greatest sculptors and painters. His best known works include the statues *David*, *Moses*, *Christ the Redeemer*, and the ceiling of the Sistine Chapel in Rome. (See Da Vinci, Leonardo).

MICHELSON, ALBERT ABRAHAM (1852 – 1931)

A Jewish-American physicist. In 1883, he invented the "interferometer" to discover the effect the movement of Earth around the sun (about 30 Km/sec) has on the speed of light. His famous experiment with Morley in 1887 showed that there is no such effect. The last experiment to measure the speed of light ended in 1935 (after his death), and was conducted by Francis G. Pease and Fred Pearson. The statement that the speed of light is constant and does not change with every frame of reference became Einstein's starting point when he developed his "Special Theory of Relativity" in 1905.

He was awarded the Nobel Prize in Physics in 1907. (See: Einstein, Albert).

NEWTON, (SIR) ISAAC (1642 – 1727)

An Englishman, and one of the greatest scientists of all times. His research started the period of classic mechanics. He invented differential and integral calculus (separately from Leibniz), the "law of universal gravitation," and discovered that white light is actually a mixture of spectral colors when he was in his twenties. In his book, *Mathematical Principles of Natural Philosophy*, he drew the laws of motion from several basic assumptions, including the concepts of "absolute space and time." He developed the "Law of Universal Gravitation" and its uses regarding the motion of planets, comets, high tide and low tide. His three central laws of mechanics are still valid today: "The Principle of Inertia," "The Principle of Force" (F=ma), and "The Principle of Equal Action and Reaction." He assumed that force, space and time had independent entities, and that this system serves as the platform for the activity of matter and motion. (See Leibniz, Gottfried Wilhelm).

OPPENHEIMER, JULIUS ROBERT (1904 – 1967)

An American physicist, born in Germany. Headed the Manhattan Project that developed the American atom bomb in World War Two. Towards the end of the war (in 1945) two such bombs were dropped on the Japanese cities of Hiroshima and Nagasaki, leading to the surrender of Japan. It shortened the war by costing hundreds of thousands of civilians who died, injured or affected by the radiation. He worked to enhance international supervision of atomic energy and objected to the production of a hydrogen bomb.

PASTEUR, LOUIS (1822 – 1895)

A French chemist. Founder of microbiology and one of the greatest scientists of the nineteenth century. He proved the existence and found the properties of bacteria as the causes of disease. He was the first to develop vaccinations. (Boiling milk to destroy the bacteria in it, is named after him). In his last 30 years of life, as his scientific and spiritual creativity was at its peak, half his body was paralyzed, as a result of a brain hemorrhage. He had a rich imagination, unusual ability to generalize and an almost "prophetic" intuition.

PAULI, WOLFGANG ERNST (1900 – 1958)

An Austrian physicist. He discovered the principle (the exclusion principle) that would then be named after him, which explains the structure of the atom shell, as well as the periodic table (*created by* Mendeleev). A commentator, and in-depth critic of new physics. Awarded the Nobel Prize in Physics in 1945.

PENZIAS, ARNO ALLAN; AND WILSON, ROBERT WOODROW

American astrophysicists. Their work and research on "Radio Galaxies" through radio telescopes, led to the discovery of "cosmic microwave radiation." Supporters of the Big Bang Theory claim that this weak background radiation is the remains of a very hot radiation that existed during "the creation of the universe" by The Big Bang.

PLANCK, MAX KARL ERNST LUDWIG (1858 – 1947)

A German physicist, founder of Quantum Theory. He formulated the concept of "energy quanta" (multiplication of "action quantum" by the frequency of radiation) as a result of his research on black-body radiation. He published important papers in the field of scientific recognition. He rejected the positivist approach in science. Awarded the Nobel Prize in Physics in 1918. (See Heisenberg, Werner).

PLATO (427 – 347 B.C.E.)

A Greek philosopher, and founder of 'The Academy' in Athens when he was 41. The academy served as the center for his philosophical ideas, until it closed in 529 AD. His main cosmological essay was "Timaeus," but many of his ideas about science, and especially about math and astronomy are included in other essays, especially in The Republic. He was the teacher of Aristotle.

POPPER, (SIR) *KARL RAIMUND (1902 – 1994)*

An Austrian philosopher. He rejected the direction of "Logical Positivism" to distinguish between meaningful statements and meaningless statements, by using the "Criterion of

Justification." Instead, he proposed "the Falsification Criterion:" "only falsifiable theories are scientific and meaningful." His books include *The Open Society and Its Enemies, Objective Knowledge,* and *An Evolutionary Approach,* among others.

PTOLEMY, CLAUDIUS (PTOLEMY) (ABOUT 85 – 165 CE)

A Greek astronomer and scientist, who worked in Alexandria. His essay *The Syntaxis Mathematica* or *Almagest* (Almagestum in Latin) consists of thirteen sections, called books, and is the most comprehensive and detailed essay to develop the Geocentric Theory. This outlook was universally accepted until the publication of Copernicus's book.

RUTHERFORD, ERNEST (1871 – 1937)

An English physicist. Founder of the "New Radioactive Theory" with Frederick Soddy. His research on the scattering of alpha rays led to the creation of his "Atomic Theory," which describes an atom, composed of a positive nucleus, which is surrounded by negative electrons. Awarded the Nobel Prize in Chemistry in 1908.

STEINER, RUDOLF (1861 – 1925)

An Austrian philosopher, scientist, researcher and educator. Founder of the anthroposophical movement. His multi-faceted contribution led to a turning point in the approach to philosophy, science, holistic medicine, education, religion, economics, history, agriculture, art, mysticism and more. His books include *Homage to Pythagoras: Rediscovering Sacred Science, Man and the World of Stars* and *Human and Cosmic Thought.*

SAMBURSKY, SAMUEL (1900 – 1990)

Born in Königsberg, Germany, Kant's city. A mathematician, physicist and philosopher. The student of Max Planck and well-known physicists. In 1920, he met Einstein in Berlin, for his famous lecture *Geometry and Experience*. He was a friend of Israel's first prime-minister, David Ben Gurion. He founded and headed the Racah Institute of Physics at The Hebrew University in Jerusalem. He established the Israeli Scientific Council, which would later become The National Council for Research and Development. He was a member of the Israeli National Academy of Science and an honorary professor at the universities of Heidelberg and Oxford. He won the Rothschild Prize and the Israel Prize. His books include *The Physical World of the Greeks, Physics of the Stoics, The Physical Thought of Late Antiquity* and the anthology: *The evolution of Physical Thought: From the Presocratics to the Quantum Physicists.*

WELLS, HERBERT GEORGE (H. G. WELLS) (1866 – 1946)

An English writer, historian, utopian, and early science fiction writer. His famous books include *War of the Worlds, The Invisible Man, The Time Machine, The Outline of History, and The First Men in the Moon.*

WHEELER, JOHN ARCHIBALD (1911 – 2008)

An American theoretical physicist. He is best known for linking the term "black hole" to objects with gravitational collapse, already predicted early in the twentieth century. Author of: eight books and many scientific papers. A professor at the universities of: Texas and Princeton and advisor of the physicists Hawking, Thorne, Bekenstein and others. Received honorary degrees from eighteen different institutions. (See: Hawking, Stephen and Bekenstein, Jacob).

ZWICKY, FRITZ (1898 – 1974)

A Swiss American astronomer, and astrophysicist. He researched galaxy clusters and found that in order for the galaxies to avoid "disintegrating " in space, a mass of up to ten times larger than the observable mass is needed, to create the required "attraction force." For a long time, this mass was termed "the missing mass." Today, one tends to assume that the mass "exists but is invisible," or "glows," and thus is termed "The Dark Mass," or in his words, "*Dunkle Materie.*" Later research has shown that the further away the galaxy cluster is from us, the higher this ratio rises, and it may even reach a ratio of 100-300 to 1 of "a missing mass" versus "an observable mass."

GLOSSARY[35]

Absolute Zero: The zero point for measuring a temperature, according to Kelvin's scale. At this temperature, all motion stops. The absolute zero is taken as −273.15° on the Celsius scale.

Acceleration: A change of speed in a unit of time. "Positive Acceleration", or Speeding up.

Amplitude: The vertical distance between the low point and the high point of a wave.

Anthology: A collection of selected literary works from different authors.

Astronaut: An American term for a man who flies or is in a spaceship or rocket in outer space. Russians call such a person a cosmonaut ("Star sailor," sailor of the universe).

Atmospheric Pressure: The pressure exerted by the atmosphere (air) on all the bodies within it. Under standard conditions, this pressure equals the weight of an almost 10 meter of water column, or a mercury column of 76 centimeters height over the area of one squared centimeter.

Atom: The basic unit of matter. It is commonly thought that the atom is composed of a miniscule nucleus, which contains protons and neutrons, around which electrons rotate in orbits. Its name is

---

35 Sources: *A Lexicon for Exact Sciences*, by Pinchas Avivi; *The Evolution of Physical Thought: From the Presocratics to the Quantum Physicists: An Anthology*, by Shmuel Sambursky; *A Brief History of Time*, by Stephen Hawking; *Encyclopaedia Hebraica*, the magazine *Galileo*, search engines, websites, online libraries and encyclopedias, the author and the editors of this book.

derived from Greek: "A-tomos," indivisible.

Atom Bomb: A bomb that uses the energy released from the fission of atomic nuclei. Fission materials include Uranium 235, Plutonium or other radioactive elements. The bomb is composed of lumps of fissile material, each of which is smaller than "the critical mass." By rapidly joining them together (through an external force) to become "a critical mass," a powerful, unsupervised chain reaction takes place, which causes the release of a vast amount of energy, destruction and radioactive pollution.

Atom Nucleus: The atom nucleus probably consists of protons, neutrons and many other particles, some of which have been validated in different experiments (See also "The Standard Model of the Atom").

Basalt: Rock formed from the rapid cooling of basaltic lava.

Basic Software: Likening the human brain to a computer, which is operated with a basic software, such as "an operating software," which must always operate according to the software instructions, and cannot operate in another way.

The Big Bang: A theory that describes the creation of the universe as a massive burst of concentrated energy, in space and time, from which the "expanding universe" was created.

The Big Crunch: A theory that assumes that the universe will end as a result of "complete gravitational collapse" of all masses in the universe to one spot, of infinite temperature and density. This could happen after the energy of the Big Bang subsides to a point where gravitation would overcome the acceleration of the Big Bang.

A Billion: A thousand million: $1,000 \times 1,000,000 = 1,000,000,000$

Black Hole: A region of space-time exhibiting such strong gravitational effects that nothing—not even light—can escape from it. There is still a controversy regarding the existence of black holes. Modern theories suggest that black holes emit energy, and have increasingly growing entropies.

Bosons: Particles that carry "the weak force" in the atom nucleus, according to "the Standard Model of the Atom," which is largely accepted today.

Braille: A tactile writing system used by people who are blind or visually impaired. The writing is made of signs that are embossed on the page or the board, and which represent a letter, a digit, a word etc. The readers can feel them, identify them and "read" with their hands.

Brownian motion: The erratic random movement of microscopic particles of matter in a fluid or gas. The motion is a result of continuous bombardment from molecules of the surrounding medium. Brownian motion was researched by the Scottish botanist Robert Brown in 1828, and named after him.

Cavor: Mr. Cavor, a fictional character in H. G. Wells, book: *The First Men in the Moon*, invented material that can block the attraction force of gravity.

Cavorite: The fictional material invented by Mr. Cavor.

CCD: An electronic camera based on a grid where each photoelectric cell receives light at a given intensity and translates the intensity of the light into an electronic signal.

CD player: An electronic device that plays audio compact discs, which contain recordings of audio material such as music.

Chain Reaction: A chain reaction that occurs when the fission of an atomic nucleus leads to the emission of neutrons, which collide and spontaneously split additional atomic nuclei. The process repeats itself, so that the number of split atomic nuclei rapidly rises, whilst emitting a great amount of energy.

Chaos: In mythology and in folktales: an infinite abyss, the darkness that existed until the creation of the world. In Greek mythology: an ancient, shapeless matter, from which the world was created through the powers of Eros, the god of love. Nowadays: a new science, which deals with the laws of complex systems, and explains phenomena such as the weather, different groups in the population, the flow of liquids and more.

"Chaos: When the present determines the future, but the approximate present does not approximately determine the future."

Classical Mechanics: According to Newton, in his book *Mathematical Principles of Natural Philosophy*, this is the theory that explains the laws of physics and mechanics, based on bodies, forces and motions. Classical Mechanics dominated scientific thought until the development of Einstein's special and general theories of relativity (1905, 1915).

Cosmic Background Radiation: A weak, homogeneous radiation in the field of microwaves. It was discovered by the astrophysicists **Arno Penzias** and **Robert Wilson** in 1965 through a radio telescope. Supporters of the Big Bang Theory believe that this weak background radiation is the remains of the powerfully hot radiation that existed during the "Creation of the universe" by the so-called **Big Bang**.

Critical: Borderline, vital, essential. The minimum required in order to start a chain reaction.

Critical Mass: The minimum amount of "fissile material" needed, so that the neutrons that are released from the fission of the atoms cause a spontaneous nuclear chain reaction. (See Chain Reaction).

254 | *Attraction – Secrets of Gravity*

Data Processing: A term borrowed from a computerized action. The way in which, a computer conducts actions, through using its software.

Deceleration: A reduction of speed in a unit of time. "Negative Acceleration, Or, slowing down.

Diaphragm: A thin membrane or partition that could be moved by gas or liquid on one side, and cause gas or liquid on its other side to move accordingly. Its motion may be translated into acoustical, mechanical, electrical, electronic and other signals (there is also a diaphragm in anatomy).

Training shell: A cannon shell without any explosive material, which does not explode when it hits the target – it serves for practice in the army.

Electric Charge: According to the Standard Model of the Atom, it is commonly thought that some of the elementary particles carry an electric charge: An electron shell carries "a negative charge" and the protons in the nucleus carry "a positive charge."

Exact Sciences: A term that describes sciences, such as mathematics, physics and chemistry.

Electric Pulse: An electric current or charge that flows at a very short period of time.

Electrical Cable: (or a Power Cable) A cable used to carry electrical power.

Electrons: Particles, charged with a negative electric charge, which usually move around the atom nucleus, in regular orbit routes. According to the hypothesis of the standard model of the atom, an electron is a particle with low mass "lepton" (light). It is assumed that electricity is a flow of electrons, which passes through, or on the surface of an electric conductor.

Elementary Particles: The elementary components of the atom, according to the standard model. These are divided into two categories: "matter particles" and "force particles." Each particle has its anti-particles. (See The Standard Model of the Atom).

Energy: A physical term with many uses. The units match units of work. According to Einstein, the energy of a body equals its mass times the speed of light squared: $E = M C^2$

Entropy: A measure of the "disorder" in the system. It is the heart of the second law of thermodynamics: The temperature (energy) of a closed system will decrease in time. The process is irreversible, and has a statistical, probabilistic nature. Raising the temperature (and reducing the entropy) will only be possible through the investment of work. In the overall balance, entropy will always grow. In this way, a direction is given to "the arrow of time:" from the past to the future, from small entropy to larger entropy.

Epistemology: A branch in philosophy *concerned with the theory of* knowledge. (Derived, from the ancient Greek *epistēmē*, meaning "justified true belief").

"It is the mark of an educated man to look for precision in each class of things just so far as the nature of the subject admits." (Aristotle)

Event Horizon: The boundaries of a black hole. The boundary around a black hole from which, even radiation cannot escape.

Fissionable: Matter, usually one of the heavy radioactive elements, capable of sustaining a nuclear fission chain reaction, which would cause its atoms to split into lighter atoms. During this process, energy is released and free neutrons are emitted.

Force: According to Newton, in his book *Mathematical Principles of Natural Philosophy*: Has an independent entity, causes motion or acceleration and may change the state of inertia of a body. In his "classical" mathematical formulation, the force equals the body mass times its acceleration (F = ma).

Fracticles: Elementary particles which are smaller by many orders of magnitudes than any known particle. They move in straight line at the speed of light. Fracticles have not been discovered yet. They appear in this book, and they form the basis of the "Repulsion Theory," instead of the "Attraction Theory."

Galaxy: A large cluster of stars, which consists of millions or billions of stars (the solar system and Earth belong to "The Milky Way Galaxy").

Gamma Radiation: Electromagnetic radiation, at a wavelength that is shorter than X-ray radiation.

General Relativity (1915): A theory that explains gravity as an outcome of "curved space-time," which forces masses to move towards each other in the same space.

Gluons: Assumed particles that 'glue' elementary particles together.

Gravitational Waves: A theoretical concept on the expansion of gravity in gravitational fields in space, similarly to other wave phenomena.

Graviton: Scientists believe that the graviton is an elementary particle that has not yet been discovered, which "carries the force of gravity."

Gravity: A quantitative description of the force that causes bodies to move towards or fall towards each other, in direct proportion to their masses and inversely proportional to the squared distance between them. The term offers an explanation to the concept of weight. The Theory of Gravity was proposed by Newton in his book *Mathematical Principles of Natural Philosophy*.

Ferromagnetic Material (from "ferrum" in Latin, meaning iron): A material that displays significant magnetic activity. Metals such as iron, cobalt, nickel, gadolinium and their alloys display significant magnetic properties, which surpass the magnetic properties of most materials.

Horse collar: Part of the harness of a horse, which is located around the neck. The straps attach to it, with which the carriage moves behind the horse.

Inertia: A basic concept in mechanics. A property of a body, by which, it resists changes in its state of motion. The mass of a body serves as a quantitative measure of its inertia. This is Newton's first law: "An object at rest stays at rest and an object in motion stays in motion with the same speed and in the same direction unless acted upon by an unbalanced force."

Infrared: Electromagnetic radiation with a wavelength that is longer than that of the red light that is visible to the human eye.

Input: The computer receives data that is entered into it, through typing or information from different sensors, such as magnetic media etc.

Kph or km/h: A speed unit expressed by kilometers per hour.

Laser: A device that emits coherent and almost parallel rays of light, unlike an incandescent light bulb or another source of light, which emit light in all directions at more or less equal intensity.

Lava: Hot, molten and primordial magma, **expelled by a** volcano **during an** eruption.

Light Year: A cosmological and astronomical unit of length (not a unit of time) to measure and mark cosmic distances. This is the distance that light passes in one year. It equals approximately 299,792.5 kilometers per second (the speed of light) times 3600 seconds per hour, times 24 hours per day, times 365 days per year - approximately 9,454,256,280,000 kilometers.

Mass: A physical term used to measure the amount of matter in an object. It is a measure of an object's resistance to acceleration (a change in its state of motion) when a force is applied. According to Einstein, mass equals the maximal amount of energy of an object divided by the speed of light squared. The more the speed of an object grows (and may reaches an infinite value at the speed of light), so will its mass grow accordingly.

Matter: Anything that is directly perceivable through the senses.

Melting Furnace: A melting oven, built for heating, and melting metals.

Mercury: The closest planet to the sun.

Micro waves: Radiation of the wave length of approximately 1 cm.

Micron: A small length unit, which equals a thousandth of a millimeter or a millionth of a meter.

Missing Mass: In order for galaxies not to " disintegrate" in space, it is assumed that a mass, which is more than ten times larger than the observable mass, is required in order to create the needed "force of gravity." For a long time, this mass was termed "The Missing Mass." Today, it is assumed that the mass "exists but is invisible," and thus is termed "The Dark Mass." (See also Zwicky, Fritz).

Molecules: A group of two or more atoms linked together by chemical bonds.

Momentum: In "classical" mechanics: The mass of the object multiplied by its speed.

Motion: The movement of an object or objects from place to place.

Net Force: A physical term from the field of vectors. It describes the sum of all forces and their direction.

Neutron: The neutron is a subatomic particle, symbol n or N^0, with no net electric charge and a mass slightly larger than that of a proton.

Nova: A nova (plural *novae* or *novas*) is a cataclysmic nuclear explosion of a "white dwarf," which causes a sudden brightening of the star for a limited period of time.

Objects: Another name for bodies.

Observation: A concept in science that refers to a meticulous, in-depth study of a natural phenomenon, which includes keeping records and archiving them, in order to conduct research and evaluation.

Ontology: A branch of philosophy. The study of things, that are existing. From: *onto*, from the Greek "on," i.e. "being; that which is," and λογία, - logia, i.e. "science, study, theory." What exists? What are the various modes of being of the different phenomena?

Optical Telescope: An optical device, with lenses and mirrors, which gathers and focuses the lights of astronomical bodies and "magnifies" them or "brings them closer" to the viewer.

Photoelectric (cell): A unit that receives light and turns it into an electric current.

Photon: A "light particle" or "a light quantum:" a unit of light or electromagnetic energy of a defined size for light of a given wavelength.

Platinum: A very heavy metal element.

Pluto: The planet that is the furthest from the sun in "our" solar system, as far as we know today.

Proton: According to the Standard Model of the atom, a heavy particle that is situated in the atomic nucleus. It is not an elementary particle. It has a positive electric charge, and it may be composed of elementary particles called "quarks."

Quantum Gravity: A theory that unifies quantum mechanics with the general theory of relativity.

Quantum Mechanics: A theory developed out of **Max Planck's** "Quanta Principle" (discrete units), which is based on **Heisenberg's** "Uncertainty Principle." It refers to energy as being radiated and absorbed not continuously, but in discrete and defined units. It gradually arose during the discovery that a "black body" emits energy in discrete values.

Quasar: Quasi-stellar radio sources that are the most energetic and distant members of a class of objects called "active galactic nuclei." A quasar emits exceptionally large amounts of energy, may be before becoming a black hole. They are observable, in spite of their great distance, due to their immense radiation.

Radio Telescope: A device for mapping radio waves received from space. Serve to research quasars and galaxies that emit radio radiation. It is also used to map the cosmic microwave background radiation.

Radioactivity: The spontaneous radioactive decay of one nucleus into one or more other nuclei, while emitting radiation.

Redshift: Redshift happens when light or other electromagnetic radiation from an object is increased in wavelength, or shifted to the red end of the spectrum. Redshift describes the tendency of star or galaxy light to turn red, due to the star or galaxy moving further away from us. This happens as a result of the lengthening of the wavelength, due to movement away from us, similarly to the Doppler Effect. Redshift of galaxies was first discovered by the American astronomer, Edwin Hubble, at the beginning of the twentieth century.

Referee: A person appointed by an institution to examine a subject and make recommendations (e.g., in a publishing house, a person who reads manuscripts and decides whether to publish them or not).

Simultaneously: At the same time.

The Sleeping Beauty: A fairy tale. "As a result of a magic spell, all the people in the kingdom fell asleep for one hundred years. And when they awoke – they continued with their actions, unaware that they had slept for such a long time."

Something: A physical entity, which does not depend on our senses or on the existence of man, in being "the thing itself" in the ontological sense, "an external physical reality."

Space-time: The concepts of time and three-dimensional space, regarded as fused into a four-dimensional continuum, according to the Theory of Relativity.

Special Relativity (1905): A theory of physics, published by Albert Einstein. Relativity emerges from two premises: the first is that the speed of light is constant in all reference systems, and the second is that the laws of physics are the same in all frameworks. One of the surprising results arising from these assumptions is that there is no absolute space and time – since as the reference system is moving at a higher speed (the speed of light being the upper limit), dimension, the length of the system in the direction of movement, is shrinking relative to the stationary observer, and the local time advances more slowly from the perspective of the viewer.

Spin: The "self-rotation" of elementary particles in the atom (electrons).

The Standard Model of the Atom: A theoretical model of atomic structure, which divides the elementary particles into families that have particles with different properties and "anti-particles" with opposite properties. An encounter between particles and anti-particles leads to their extinction, whilst emitting energy. It is assumed that more than 24 different particles exist in a number of families, with sub-divisions in each family.

Subjectivity: A point of view, perspective, or perception which is personal to the viewer or the thinker.

Supernatural: Magical, as a result of a spell, somewhat mystical.

Supernova: A very bright star. The brilliant point of light is believed to be the explosion of a star that has reached the end of its life. It is known as a supernova. Supernovas can briefly outshine entire galaxies and radiate more energy than our sun will in its entire lifetime.

Switch: A device for making and breaking the connection in an electric circuit.

Theory of Relativity: Developed and formulated by **Albert Einstein**.

Tug of War: A group sport, two groups pulling a thick rope in opposite directions. It was popular in the early days of the State of Israel and during its first few years.

Ultra Violet: A short wavelength, beyond the wavelengths that are detectable by the human eye.

Universe: A concept that describes all physical bodies that are detectable or assumed and the space containing them.

Vacuum: "Sub pressure," "suction," gaseous pressure in a vessel, which is lower than the ambient pressure around it.

Voltage Source: In electricity: any source that creates a difference between electrical potentials (such as an electric battery or a generator), which, when plugged in through an "electric circuit," would cause an electric current to "flow."

X Rays: Electromagnetic radiation at a very short wavelength: between 5×10^{-7} – 5×10^{-10} cm. This radiation is used in medicine for taking x-ray photographs. It reaches us from space, possibly from black holes.

Sources of Citations
& References

Anaxagoras / *Principle of Predominance*, from H. Diels and W. Kranz, *Die Fragmente der Vorsokratiker*, Sixth Edition, Berlin.

Aristotle / *A blog by Paul Nadal/ What Is Time?: On Aristotle's Definition of Time in Physics Book IV.*

Bekenstein, Jacob / *Black Holes and Everyday Physics. Physics Department,* Ben Gurion University of the Negev, Beer-Sheva 84120, Israel.

Bohr, Niels / *On Atoms, And Human Knowledge*, from the paper *Atomic Physics and Human Knowledge,* London, 1958.

Bruno, Giordano / *de Bibliotheca: Telematics Library Classic Italian Lieterature of the Infinite Universe and Worlds of Giordano Bruno First Dialogue.*

Da Vinci, Leonardo / From: *How to Think Like Leonardo Da Vinci,* by Michael J. Gelb.

Democritus / From *Die Fragmente Der Vorsokratiker*, by H. Diels, Sixth Edition, Berlin 1951.

Descartes, René / *Meditations on First Philosophy.*

Einstein, Albert / From Yossi Lev, Metaphor, *A Forgotten Language,* Galileo, Scientific Magazine, July-August 1999.

Epicurus / *On Atoms and the Void*, from *Letters to Herodotus and to Menoeceus.*

Erhard, Werner / From *Werner Erhard, The Transformation of a Man: The Founding of EST*, by W.W. Bartley.

Faraday, Michael / *Experimental Researches in Electricity, Gravitation and the Concept of the Force field, 1839 – 1855, London Press*.

Feynman, Richard P. *The Meaning of It All: Thoughts of a Citizen Scientist*. Reading Mass: Addison-Wesley 1998.

Galilei, Galileo / *Discourses and Mathematical Demonstrations Relating to Two New Sciences*, 1638.

Galilei, Galileo / Dialogues Concerning Two New Sciences. Translated from the Italian and Latin into English by Henry Crew and Alfonso de Salvio.

Gilbert, William / *De Magnet*. London Press, 1900, Volume 1, Chapter 17.

Hawking, Stephen / *A Brief History of Time*. 1992.

Hawking, Stephen / lecture: *Into a Black Hole*.

Heisenberg, Werner / *Physical Principles of the Quantum Theory*.

Hertz, Heinrich / *www.azquotes.com/author/22801-Heinrich_Hertz*

Kant, Immanuel. *The Metaphysics of Morals (1797)*. Translated by Mary Gregor. Cambridge University Press, 1996 Adapted with Extracts Jeffrey W. Bulger.

Leibniz, Gottfried Wilhelm. *G. W. Leibniz: Philosophical Essays*. R. Ariew and D. Garber, eds., Indianapolis: Hackett, 1989.

Maxwell, James Clerk. *Scientific Papers, (1861 – 65) Volume One*. from the paper *A Dynamical Theory of the Electromagnetic Field*. Cambridge University Press, 1890.

Maxwell, James Clerk: *Scientific Papers, On Physical Lines of Force*, 1861.

Newton, Isaac / *Mathematical Principles of Natural Philosophy*, 1687, from the third law "Action and Reaction."

Newton, Isaac / *Mathematical Principles of Natural Philosophy*, Second Edition, with preface by Roger Cotes, 1713.

Popper, Karl / *From: The Limits of Reason: Thought, Science and Religion; Yeshayahu Leibowitz and Joseph Agassi in Conversation, by Chemi Ben Noon, Keter Publishers, 1997.*

Sambursky, Shmuel. *A General Introduction to the Anthology: The Evolution of Physical Thought: From the Pre-Socratic to the Quantum Physicists.* Jerusalem: Bialik Institute Press, 1972.

Thorne, Kip / APPLICATIONS OF CLASSICAL PHYSICS: Roger D. Blandford and Kip S. Thorne/

Wheeler, John
www.brainyquote.com/quotes/quotes/j/johnarchib108110.html